About this book

I0505675

This book contains ecological field studies techniques, that can be conducted in wild nature by non-professional researchers - school and university students together with their teachers, single beginning investigators, families, amateurs of all ages.

The whole book "Field Studies Techniques" includes 5 parts/series corresponding to a specific field sciences (Geography, Botany, Zoology, Hydrobiology and Bioindication) with totally 40 environmental study lessons (see below) covered a wide variety of activities in nature which can be arranged in different seasons of the year. Each series contains techniques focusing on practical skills which can be applied in the field research.

This book promotes a wide variety of outcomes which correspond to established educational standards in many countries. The ecological field study activities address content standards in the areas of earth science, life science, biology and ecology. Intellectual skill development includes questioning, data collection, analysis and drawing conclusions.

This book promotes understanding of ecosystems and the protection of the environment through the training of teachers in specific field study techniques, the education of young people in ecology concepts and issues and the sharing of ecological study results between colleagues. The primary goal of this work is connecting students and teachers, addressing important environmental issues and promoting environmental interest, knowledge and values.

This book is addressed to the middle and secondary level science teachers and students, and for all those who would like to investigate local wild nature, to share ecological and cultural information and work together to help create a better environment.

This specific book "Field Studies Techniques. **Part 5: Bioindication and Nature Monitoring**" is the fifth in the series. Practical skills include biological methods of the assessment of air pollution, vital state of the forest, ecological features of the meadows, environmental state of rivers, as well as general evaluation of a human impact on an area and complex integrated study on landscape profile.

The techniques described in this book, were written by Alexander Bogolyubov (1996-2002), translated into English by Tatiana Tatarinova (2002) and edited by Michael Brody (2003).

The list of all field study lessons:

I. Geography and Landscape Sciences:

Orienteering in the forest

Simple "Eye" Survey of the Field Study Site

Mapping Forest Vegetation

Procedure of the Geological Exposure Description

Studying Minerals and Rocks in Your Area

Plotting a Profile of a River Valley Slope

Simple Procedure of Soil Description

Integrated Study Based on Landscape Profile

Complex Comparative Description of Small Rivers and Streams

Study of Snow Cover Profile

Making a Campfire

II. Botany:

Study of Species Composition and Number of Fungi

Making a Herbarium

The Study of Plants in Your Local Environment

Study of the Vertical Structure of a Forest

Mapping Forest Vegetation

Green Plants Under Snow

Study of the Ecology of Early Flowering Plants

Phenology of Plant Florescence

Assessment of Ecological Features of Meadows on the Base of Vegetation Cover

Assessment of the Vital State of Coniferous Underbrush

Study of Growth Dynamics of Trees Based on Annual Rings

Assessment of the Vital State of a Forest Based on Pine-tree Analysis

Assessment of Environmental State of the Forest Based on Leaves' Asymmetry

III. Zoology:

Study of Forest Invertebrates in the Forest Litter and Wood

Study of Forest Invertebrates in the Grass Layer, Tree Crowns and Air

The Study of Water Invertebrates in a Local River and Assessment of Its Environmental State

Studies of Species Composition and Abundance of Amphibians

Let's Help Birds! (Making Feeders and Nesting Boxes)

Study of Species Composition and Census of Birds Using the Line Transect-counting Method

If you have any questions or need an advice on the lessons below, write to ecosystema1994@yandex.ru.

_____

* **Bold** = important text phrases, ***Bold/Italic*** = key vocabulary, *Italic* = examples and specific illustrations.

Assessment of air pollution
by lichen indication method

This manual contains procedures for the application of **bioindication** research using lichens for the assessment of **environmental conditions**. The focus is on methods of quantitative evaluation of **lichen** species in biological monitoring based on the use of **poleotolerance** classes and lichen **bioindication indexes**.

Introduction

The response of biological objects to pollutants is significant for investigations of environmental **pollution** caused by industrial means. A system of observations of the response of biological objects to the impact of pollutants is called *biological monitoring*.

Biological monitoring includes observation, assessment and forecast of changes in the state of ecosystems and their elements caused by *anthropogenic* factors.

Lichens have been chosen as one of the main objects of global biological monitoring. Lichens represent a peculiar group of spore plants consisting of two components – a *fungus* and a single-celled or, rarely, filamentous *algae*, which live together and function as one organism. The fungus performs the functions of reproduction and feeding at the substratum, whereas the algae photosynthesize.

Lichens are very sensitive to nature and composition of the substratum; their growth is dependent on *microclimatic* conditions and composition of air.

Due to the extraordinary longevity of lichens, they can be used for dating the age of various objects over the range of several decades to thousands of years, based on measurements of their *thallus*.

Lichens were chosen as an object of global monitoring due to the fact that they are widespread on earth and their response to external influence is very strong, whereas their own variability is small and extremely slow compared to other organisms.

Epiphytic lichens (or epiphytes), i.e. lichens growing on bark of trees, are characterized with the most sensitivity among all ecological groups of lichens. Study of these species in large cities of the world revealed a number of general patterns: the more industrialized the city is, the fewer species of lichens are found; the less the total area of the tree trunk is covered with lichens, the lower the vitality of the lichens.

It has been discovered that when the level of **air pollution increases**, first *fruticose* lichens disappear, then *foliaceous* lichens, and the last ones to disappear are *crustose* (cork-forming) forms of lichens. Composition of lichen flora in different areas of a city (in the center, in industrialized districts, in parks and in the outlying districts) turned out to vary so great that researchers began to use lichens as indicators of air pollution.

A Swedish scientist, R.Sernander, conducted one of the **first studies in this field** in 1926. He pointed out a "lichen desert" in Stockholm (a center of the city and industrialized areas with high level of air pollution) where lichens were almost completely absent. He also noted a "competition zone" (districts of the city with average level of air pollution) where the number of lichens was poor, and a "normal zone" (outlying districts of the city) where many species of lichens were found.

Recently it was shown that among all components of polluted air, **sulfur dioxide (SO$_2$)** has the strongest **negative influence on lichens**. It was experimentally proven that a concentration of sulfur dioxide equal to 0.03 – 0.1 mg/m^3 (30-100 microgram/cubic meter) had an impact on many species of lichens. At that concentration, brown spots appear in **chloroplasts** of alga cells and degradation of chlorophyll begins. A concentration of sulfur dioxide equal to 0.5mg/cubic meter is harmful (even destructive) for all lichen species found in natural landscapes. There is a group of **poleotolerant** species (resistant to pollution), that can exist in rather polluted air.

In addition to sulfur dioxide, **other pollutants** including nitrogen oxides (NO, NO$_2$), carbon oxides (CO, CO$_2$) and fluorine compounds also have an adverse effect on lichens. Moreover, **microclimatic conditions** in cities have changed greatly from their natural origins: cities are "drier" in cities, compared to natural landscapes (approximately 5% drier) and 1-3% warmer. Cities are also characterized by poorer natural illumination.

With this information in mind, lichens serve as integral indicators of the state of the environment. The presence of Lichens in general indicates that *abiotic* factors are favorable for life.

Most chemical compounds, having a negative impact on the flora of lichens, represent the main chemical elements and substances found in the **emissions of most industrial** objects. This information allows us to use lichens as indicators of the *anthropogenic* load.

All the above-mentioned factors helped to determine the application of lichens and lichen bioindication in the system of global environmental monitoring. The present lesson is also based on attempts to use lichens as bioindicators of the state of environment. This lesson is aimed at assessment of **spatial differences in air pollution based on the use of lichens**.

The following **items will be required** for the lesson: transparent overlay grid, measuring tapes with millimeter points (1 meter long) and a compass.

General procedure of the research

This research, like many other studies devoted to environmental monitoring, is rather difficult and has many different aspects. In this section, we will provide a **general description** of the research as a whole, and then the following sections will focus on **specific rules** of lichenoindication studies, techniques of material collection, and methods of data processing of field measurements.

The work should begin by familiarizing students with the concept of environmental monitoring and bioindication, the biological features of lichens as bioindicators as well as the main methods of bioindication.

The **introductory (theoretical) part of the lesson** should be

devoted to familiarizing students with objects of study – lichens. It is advised to have a collection of lichens at the school or environmental field center; however, the collection cannot replace observing live specimens of these unique and beautiful organisms. In order to show the main species of lichens to students, it is recommended to have an excursion to one of the *biotopes* that is rich in lichens, or to the site, where a rich diversity of lichen species can be found. For example, in our Field Center "Ecosystem" in Russia, we keep a piece of old plank fence especially for that purpose. Up to 15 species of lichens can be found on a three-meter-long section of the fence, which represents almost the entire diversity of lichens in the given area.

The next stage of the lesson is familiarization with a procedure of **evaluating lichen population**. Demonstration measurements of lichen numbers are conducted by two major methods on any trunk covered with lichens, with the help of an **overlay grid** (transparency) and with a **measuring tape** (see below).

Independent research starts with the selection of sites for studies. It is recommended to find four or five sites, preferably located along one line at specified distances away from a potential source of pollution in your area.

The **distance between sites** depends on the impact (power) of the source of pollution. If it is a large human settlement with industrial plants and numerous vehicles, then distances between sites can vary up to 1 km (in this case, the farthest site will be 5 km away from the town). If it is a highway, then distances can vary from 20 to 100 meters (depending on intensity of traffic).

For the purposes of this study, it is not necessary to **mark sites;** however, special attention should be paid to the selection of sites for study (see below "Main rules of lichenoindication research").

Students are **divided into teams** of two or three in order to carry out field studies. Each team is given a task to investigate one of chosen sites. After all field measurements have been made, all teams return to the classroom and begin to process collected material under the direction of their teacher. They summarize all their results as one research document by preparing an article or oral report.

Methods of lichenoindication

Methods of lichenoindication are divided into two large groups: **active** lichenoindication and **passive** lichenoindication.

Active lichenoindication includes so-called **transplant methods**. Transplant methods are based on the procedure in which lichens from unpolluted areas are transplanted into a studied district or when sections of tree bark covered with lichens are cut off and transferred onto posts or other facilities located in polluted districts. Their response is investigated by periodical measurements or photographs.

Another (purely experimental) approach consists in the transfer and **investigation of lichens in the laboratory**, where they are exposed to different concentrations of contaminants. One of the first signs of damage to lichens is the decrease of thallus thickness as well as *chlorosis* due to destruction of chloroplasts. The reproductive structures of lichens change or stop their development. The level of pollution can be evaluated according to rate that lichens die out.

Lichens growing on withered branches are often used for transplantation. A branch from a clean area is transferred into a studied district and placed in conditions closest to their original habitat's illumination and moisture levels and keeping its spatial orientation.

The main method of passive lichenoindication is the observation of changes in relative numbers of lichens. In order to do this, it is necessary to measure projective cover of lichens at permanent or variable test-sites and to calculate the average values of projective cover for the studied area. The same measurements of projective cover are carried out at similar sites or at the same sites in a certain period of time. It is possible to discern an increase or decrease of pollution in an area or time on the basis of the changes in total projective cover and projective cover of separate lichen species, using scales of lichen sensitivity and special indexes.

Test-sites can be both permanent (long-term) and used in the course of several years, and variable, i.e. "used once."

Main rules of lichen indication studies

The **need for development of strict standards** in the use of lichenoindication procedures appeared when scientists started to use lichens for the purposes of environmental monitoring.

Recently it was shown that using incorrect procedures of lichenoindication misled researchers (especially beginners) with resulting erroneous results. For example, the comparison of two sites (within the city and outside of the city) of lichenoflora can be invalid when lichens are studied on the bark of *lime-trees and maples in the city*, and then on *pine-trees and birches in the forest*. There is no point in collecting such data; moreover, there is no sense in analyzing it.

The following **main rules** should be complied with when organizing lichenoindication monitoring. It is preferable to study lichens at permanent (long-term) sites and test trees in the course of a long time as opposed to single (one-time) investigation of series of test-sites. Test sites should be laid down in phytocenoses, which are homogeneous according to their composition and age of trees (it is ideally when sites are located in mono-species, even-aged forest plantations).

Biotic and abiotic environmental conditions at comparable test-sites should be equal if possible (including composition and structure of phytocenoses, form of relief, moisture and illumination levels, etc.). Test trees at test-sites should be constant, if possible, but not random. In any case, test trees at the test-sites should be of the same age without noticeable damages; they should belong to one of the major forest-forming species. When using variable test-sites (while conducting "one-time" investigations), you should use fewer than ten sites (depending on objectives of the study); whereas the number of test trees at each site should be several dozen in order to obtain a large amount of statistically reliable information.

Selection of sites and test trees

The **procedure of selecting test sites and test trees** where lichen communities will be studied is very important. We can even say that it is the foundation of lichenoindication studies. A test site is a part of the area (in a typical case – a part of the forest), where lichenoindication research is carried out and where test trees are selected.

There are **several approaches** to the procedure of test site

selection, which depend on the duration of site use, i.e. whether research is short-term ("one-time") or it is planned for the course of many years. When it is necessary to carry out a census of

lichens at **several remote sites** (for example; studying the impact of pollution caused by an object on the environment along the transect line, away from the object), test sites and test trees are chosen arbitrarily and not marked. This is referred to as "variable or transient" sites.

If monitoring is planned for the **long-term**, i.e. for the course of several years, permanent (constant) sites are chosen. They can be the same sites that are used as standard geobotanical test sites or sites for assessment of the vital state of forests. Regardless of what studies are planned, the **following rules should be followed when selecting sites**:

a) Avoid roadside trees, as other factors have great impact on their *epicover*, in contrast to trees growing far from roads

b) Avoid over-stocked forest stands with very low levels of illumination

c) Beware of pastures and meadows that were treated with pesticides or were intensively fertilized.

In both cases: during a single (one-time) study and when planning long-term observations, **test trees within the test site are chosen randomly**, i.e. in a random order regardless of the fact if lichens grow on them abundantly or there are no lichens at all.

Procedure of laying down test sites

The center of the test site is marked in the forest, for instance, a stake is driven in the ground or one of the trees is marked with paint. Twenty trees (not less than 10 trees, in any case) closest to the center of the site are chosen; the trees should belong to the same species and be of approximately the same age. No exceptions of the human factor (subjective considerations) are allowed (for example, a tree is rich in lichens or it has few lichens). These test trees are then used for a census of lichen populations.

For a single (one-time) study, trees are not marked, but when planning long-term research; trees should be marked with the help of permanent markers. Metal (aluminum, brass) plates with stamped (harrowed) numbers can be used as markers. The plates should be fastened to trees with tacks. Their presence on a trunk will not influence lichen population or the general vital state of the tree. Markers should be placed on the side of the tree facing the center of the test site, so that all marked trees are well seen from one point.

Procedure of evaluation of relative numbers of lichens

In order to evaluate numbers of lichens on trees, and in particular, their projective cover, **two main techniques are usually used**: a method of "***linear intersections***" and a method of "***overlay grid***." Both techniques give roughly the same results; however, we recommend the first one – a method of "linear intersections" – in order to unify results when performing the present educational task.

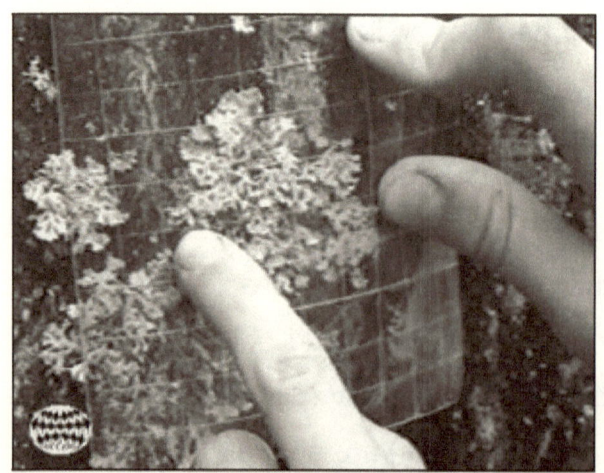

The method of "**overlay grid**" is less accurate, though it is more vivid and thus it can be used for educational purposes. On the whole, in spite of the clarity and simplicity of the technique, a shortcoming of this method of measurements is the complexity of the separate evaluation of numbers of each lichen species. For example, if there are several species of lichens at the examined section of the tree bark, then the procedure of estimation of the projective cover becomes complicated as each species should be calculated separately. In this case the investigation of one site (even if it has the size of 10 x 10 cm) can take a lot of time.

The method of "linear intersections" does not have this shortcoming; it is less vivid and requires calculations that are a little more complicated, but at the same time it is more accurate and universal.

Regardless of the chosen method of lichen census, **all**

measurements are taken at the same height – approximately 150 cm above the ground (the most important aspect is to make all measurements the same).

Before the beginning of census, **special tables should be prepared**, where the main information on location of the site of measurements as well as calculation results are recorded:

Description of the test site:

1. Date:
2. Number:
3. Location:
4. Exposition and angle of the slope:
5. Description of the phytocenosis:
6. Surnames of researchers:

Description of test trees and results of measurements:

	Lichen species:	Location of thalluses (cm):	Projective cover (%):
1. Number of the tree: 2. Species of the tree : 3. Height of the tree : 4. Length of the trunk circumference: *800 см*	1. ... 2. ... 3.	*7,1-8,5; 12,7-14,2; 30,4-32,5;* *56,4-58,8;* ...-...; ...-... ...-...	9,25 ... ...
1. Number of the tree: 2. Species of the tree : 3. Height of the tree : 4. Length of the trunk circumference:	1. ... 2.	...-...; ...-... ...-...	... ...

Unlike the "overlay grid" method, the estimation of lichen projective cover with the help of the "linear intersections" method is based on measurement of linear, but not square, parameters. **The method consists of placing a measuring tape on the trunk circumference**, and then recording all intersections of the tape measure with lichen thalluses. A common "tailor's tape" (with millimeter markings) can be used as a measuring tape.

Lichen measurement is carried out as follows:

After a test tree has been chosen, the researcher finds a point on the trunk that is 150 cm above the base of the tree on the northern side (this is determined with the help of compass). Then a measuring tape with millimeter markings is placed on the trunk so that the zero of the scale coincides with the chosen point, and numbers on the scale increase clockwise (in the direction of north to east).

After a full turn round the trunk, the tape is fixed on the tree with a pin in the zero point. **Circumference of the trunk is determined** by matching the last point on the tape with the zero point. This number represents 100% in subsequent calculations.

Then measurements are taken, moving along the tape and

registering the beginning and end of each intersection of the tape with lichen thallus (it is convenient to use a pointer – a pencil, pen, match, etc.) in order not to make a mistake. It is easier to take measurements with two people; one student will count off distances on the tape and dictate them, while the other writes down the values in their field diary.

Projective cover of lichens is calculated at home on the basis of field measurements, i.e. the ratio of the trunk section covered with lichens to the general surface of the trunk should be determined.

First, the general (total) length or extent of lichen thalluses is calculated. When the total circumference of the trunk is calculated and taken for 100%, then the projective cover of lichens is estimated (as a percentage).

Example:

Records in the field diary indicate that along the circumference of the trunk (which is 80 cm or 800 mm long) intersections of the measuring tape with lichen thalluses are registered at the following marks: *7.1-8.5 cm, 12.7-14.2 cm, 30.4-32.5 cm, 56.4-58.8 cm*. The total length (sum) of lichens is equal to 7.4 cm (1.4 + 1.5 + 2.1 + 2.4). According to the rule of proportion calculations, where 80 cm represents 100% and 7.4 cm is taken as X % (7.4/80 X 100), we can estimate the value of projective cover, which will be equal to 9.25%.

Projective cover can be estimated both for each lichen species separately and for all lichens in total; it depends on the students' knowledge and their teacher. The given lesson makes provisions for two variants of subsequent calculations: when students have to determine lichen species, and when they do not have to do it. However, projective cover should be estimated in both cases.

Projective cover according to the method of "linear intersections" (as well as when applying the "overlay grid" method) is estimated at several test trees within constant or once-used test sites. As it has been mentioned above, it is advised to examine about 20 (but not less than 10) trees.

Certain difficulties arise when taking measurements of *epiphytic* lichens, as they often represent branched twigs, which outstretch on a substratum; they can be upright or drooping. Twigs are less than 1 mm thick in most cases, so values of the lengths of their intersections with the measuring tape are systematically overstated. In some cases, several dozen fruticose lichen twigs intersecting with the measuring tape can be found on the trunk, thus, overestimation of the projective cover on the whole trunk, which is calculated as a sum of lengths of separate intersections, can reach a great value. In order to make the margin of error smaller when taking measurements, we recommend recording just the number of twigs of the given lichen that are crossed with the tape. Then the total value of projective cover of the given lichen species can be calculated on the basis of the average thickness of one twig.

Processing results of field measurements

As it has been stated above, **bioindication is based on the law of ecological individuality of species**. Different species respond to environmental factors (including anthropogenic ones) differently – each species has an individual ecological range from optimal to lethal environmental conditions. The general concept of lichen typology (classification) according to their endurance (poleotolerance, sensitivity, sensibility – all these terms are synonyms and can be found in scientific literature) concerning environmental pollution was developed in 1960s on the basis of the above statement.

Two approaches: qualitative and quantitative, are applied for assessment of environmental pollution at a certain area with the help of lichenoindication methods.

In the first case, the **level of environmental pollution of the area** is determined on the basis of **thorough scrutinizing of the species composition of lichens**. Applying data on the presence or absence of certain lichen species at the studied area and by using special tables of poleotolerance classes, which have been developed by lichenologists (see below), it is possible to determine to which conditional category the studied area belongs.

In the second case, special **lichenoindication indexes**, which take into account both the found lichen species' membership in a certain poleotolerance class and the quantitative measurement data of lichen population are used for the assessment of environmental pollution of the area.

Application of poleotolerance classes of lichens

As a result of long-term field and experimental investigations, a system of **poleotolerance classes of lichens has been developed**. Poleotolerance classes are groups in which their members give more or less similar responses to specific pollutants and their concentrations in ambient air (Table 1, below).

The comparison of species composition of lichens found within a certain area with the data of the given table will help to conditionally determine the level of total, **integrated "disturbance" of the area**, including disturbance resulting from air pollution.

Lichenoindication indexes help to determine the level of habitat disturbance more accurately and, what is more important, quantitatively. They primarily take into account the diversity of species, i.e. the number of species and numbers (populations) of different species of lichens.

Table 1. Types of habitats according to degree of impact of anthropogenic factors and occurrence of lichen species in the habitats.

Application of indexes of lichenoindication

Types of habitats	Lichen species	PC*
Natural habitats (landscapes) without tangible anthropogenic impact	Lecanactis abietina, Lobaria scrobiculata, Menegzzia terebrata, Mycoblastus sanguinarius, species of genera Pannaria, Parmeliella, the most sensitive species of the genus Usnea	I
Natural (frequently) and slightly anthropogenically transformed habitats (rarely)	*Bryoria* chalybeiformis, *Evernia* divaricata, *Cyalecta* ulmi, *Lecanora* coilocarpa, *Ochrolechia* androgyna, *Parmeliopsis* aleurites, *Ramalina* calicaris	II
Natural (frequently) and slightly anthropogenically transformed habitats (frequently)	*Bryoria* fuscescens, *Cetraria* chlorophylla, *Hypogymnia* tubulosa, *Lecidea* tenebricosa, *Opegrapha* pulicaris, *Pertusaria* pertusa, *Usnea* subfloridana	III
Natural (frequently), slightly-to frequently-anthropogenically and moderately-transformed habitats (rarely)	*Bryoria* implexa, *Cetraria* pinastri, *Graphis* scripta, *Lecanora* leptyrodes, *Lobaria* pulmonaria, *Opegrapha* diaphora, *Parmelia* subaurifera, *Parmeliopsis* ambigua, *Pertusaria* coccodes, *Pseudevernia* furfuraceae, *Usnea* filipendula	IV
Natural, anthropogenically slightly- and moderately-transformed habitats (with equal occurrence)	*Caloplaca* pyracea, *Lecania* cyrtella, *Lecanora* chlarotera, L.rugosa, L.subfuscata, L.subrugosa, *Lecidea* glomerulosa, *Parmelia* exasperata, P.olivacea, *Physcia* aipolia, *Ramalina* farinacea	V

Natural (comparatively rarely) and anthropogenically moderately-transformed (frequently) habitats	*Arthonia* radiata, *Caloplaca* aurantiaca, *Evernia* prunastri, *Hypogymnia* physodes, *Lecanora* allophana, L.carpinea, L.chlarona, L.pallida, L.symmictera, *Parmelia* acetabulum, P.subargentifera, P.exasperatula, *Pertusaria* discoidea, *Hypocenomyce* scalaris, *Ramalina* fraxinea, *Rinodina* exigua, *Usnea* hirta	VI
Moderately- (frequently) and strongly- (rarely) anthropogenically transformed habitats	*Caloplaca* vitellina, *Candelariella* vitellina, C.xanthostigma, *Lecanora* varia, *Parmelia* conspurcata, P.sulcata, P.verruculifera, *Pertusaria* amara, *Phaeophyscia* nigricans, *Phlyctis* agelaea, *Physcia* ascendens, Ph.stellaris, Ph.tenella, Physconia pulverulacea, *Xanthoria* polycarpa	VII
Anthropogenically moderately- and strongly-transformed habitats (with equal occurrence)	*Caloplaca* cerina, *Candelaria* concolor, *Phlyctis* argena, *Physconia* grisea, Ph.enteroxantha, *Ramalina* pollinaria, *Xanthoria* candelaria	VIII
Anthropogenically strongly-transformed habitats (frequently)	*Buellia* punctata, *Lecanora* expallens, *Phaeophyscia* orbicularis, *Xanthoria* parietina	IX
Greatly anthropogenically transformed habitats (occurrence and viability of lichens are low)	*Lecanora* conizaeoides, L.hageni, *Lepraria* incana, *Scoliciosporum* chlorococcum	X

* PC - Poleotolerance classes

At present there are **several dozen lichenoindication indexes developed**, which include both: indexes which take into consideration species composition of lichens; and indexes which require only estimation of species diversity (a number of species).

We will describe only the two simplest indexes here for educational purposes.

Index #1 Poleotolerance

Index of poleotolerance (IP) takes into consideration the **species composition of lichens** (i.e. in order to apply the index it is necessary to determine species) and is calculated according to the formula: $IP = \sum_{i=1}^{n} \dfrac{AiCi}{Cn}$, where: "n" is a number of species at the examined test site; Ai – poleotolerance class of i-species (from 1 to 10 cm, the right column of the table), Ci – projective cover of i-species in points, Cn – sum of projective covers of all species (in points).

Index of poleotolerance is estimated for all studied test trees at the site as an average value.

The total examined area of trunk surface when using "overlay grids" should not be less than 0.7; when using a measuring tape, not less than 20 meters in circumference.

Estimation of projective cover is made according to a 10-point scale:

Points	1	2	3	4	5	6	7	8	9	10
Cover, %	1-3	3-5	5-10	10-20	20-30	30-40	40-50	50-60	60-80	80-100

Values of IP vary in range from 1 to 10. The more the IP value is, the more polluted the air is in the corresponding habitat. IP can be equal to zero only in the case of a total lack of lichens.

Example:

The following data was obtained in the result of investigations of projective cover on 20 test trees within one test site with the help of a measuring tape:

Species # 1 – average projective cover is 15%, Species # 2 – 10%, Species # 3 – 3%, Species # 4 – 1%.

We then find corresponding values of Ci in points according to the table:

For species # 1 – 4 points, for species #2 – 3 points, for species #3 – 2 points and for species # 4 – 1 point.

Total sum of cover – Cn is equal to 4+3+2+1=10 points.

Let us assume that according to table 2, species # 1 = 6th poleotolerance class, species #2 = 7th poleotolerance class, species # 3 = 7th poleotolerance class and species # 4 = 8th poleotolerance class.

Then we calculate IP according to the formula on the basis of obtained values:

$$IP = ((4X6)/10) + ((3X7)/10) + ((2X7)/10) + ((1X8)/10) = 6.7$$

Now this index can be compared with similar indexes calculated for other test sites.

27

Values of IP correlate with annual content of SO_2 in the air:

IP	Concentration of SO_2 (mg/m³)	Conditional area
1 – 2	< 0,01	Normal
2 – 5	0,01 - 0,03	Area of low pollution
5 – 7	0,03 - 0,08	Area of moderate pollution
7 – 10	0,08 - 0,10	Area of heavy pollution
10	0,10 - 0,30	Area of crucial pollution
0	More than 0,3	Lichen desert

Index #2 Atmospheric Quality

Another method of calculation, which does not require knowledge of lichen species composition is the **Index of Atmospheric Quality** (IAQ): $IAQ = \sum_{i=1}^{n} \frac{QiCi}{10}$, where:

Qi is the ecological index of a certain i-species (or index of associativity), Ci is index of abundance of i-species, and n is a number of species.

First, IAQ is calculated for each test tree separately; then the average value for the test site as a whole is estimated.

Ecological index (index of associativity) Q characterizes a number of species, which accompany the given species at the whole test site, plus the examined species as well. In fact it is the total number of species found at the given site.

Estimation of projective cover is made according to the same 10-point scale as when calculating the index of poleotolerance.

So, **the more the projective cover of lichens, the higher the IAQ**, and, correspondingly, the cleaner the air in the given habitat. Values of IAQ can vary in range from 0 to infinity, theoretically.

Example:

Let us assume that three different species of lichens with corresponding values of projective covers of *5, 15 and 25%* are found on tree # 1. Altogether 12 lichen species are registered at the site (on all test trees). Thus, the index of associativity Q for each species is equal to 12.

Now we can estimate projective cover in points: *species # 1 = 3 points, species # 2 = 4 points and species # 3 = 5 points.*

We will make further calculations by placing the obtained results in the formula:

IAQ = ((12 X 3)/10) + ((12 X 4)/10) + ((12 X 5)/10) = 14.4

The same calculations are made for all test trees, and then we estimate the average index for the studied site as a whole.

Next, the obtained average index for the site can be compared with similar indexes, estimated for other sites.

Just as the index of poleotolerance, the index of atmospheric quality correlates with concentration of SO_2 in ambient air (according to Trass, 1985).

IAQ	Concentration of SO2, mg/m3
0 - 9	More than 0,086
10 - 24	0,086 - 0,057
25 - 39	0,057 - 0,028
40 - 54	0,028 - 0,014
More than 55	Less than 0,014

Obtained **results can be presented** in the form of a graph, where location of test sites is plotted along the horizontal axis according to chosen scale (the sites are marked according to increase of distance from the source of pollution), and indexes of air pollution (IP and IAQ) at the given sites are marked on the vertical axis:

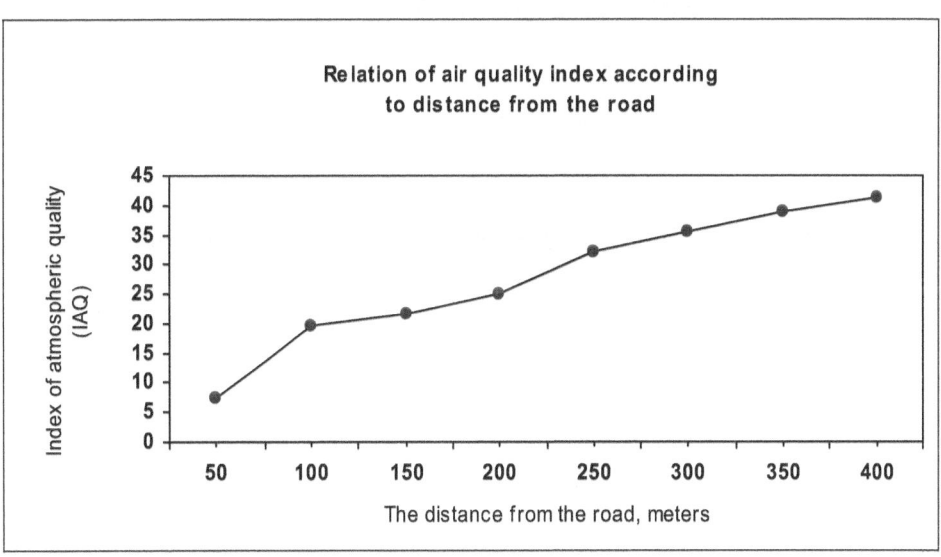

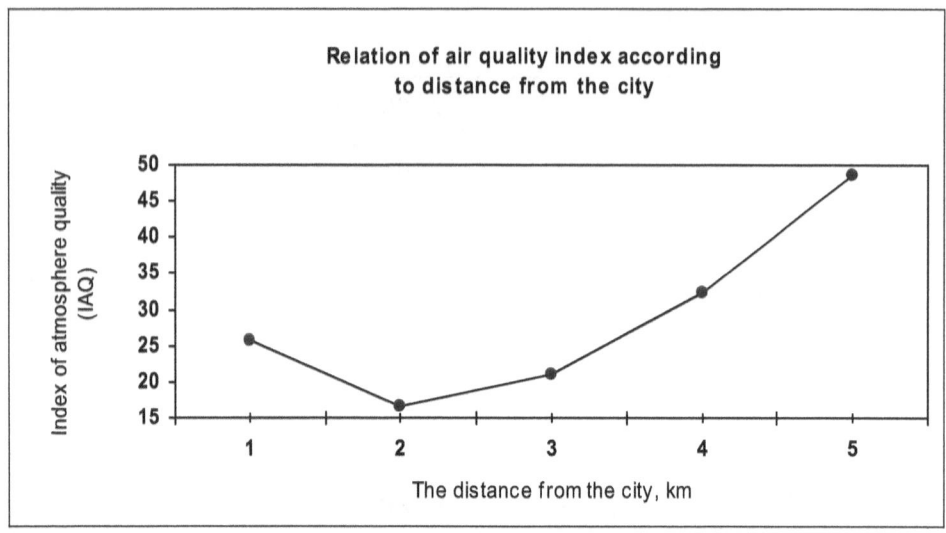

Final recommendations

The **success of lichenoindication research** depends primarily on the accuracy of design of an experiment (location of test sites), the amount of collected material and the reliability of measurements.

Let us repeat **the main rules of such lichenoindication investigations**, which are carried out within the framework of the present lesson.

Several sites should be selected for studies of lichens according to the map of the area within the vicinity of the school or environmental center. Sites should be located along the transect line, away from the potential pollution source at a distance of 300 to 1000 meters from each other (depending on the scale of the pollution source). It is advised to choose no fewer than 4-5 sites.

Even-aged forests consisting of one species are recommended for the studies of lichens, if possible. The forests should be characterized with approximately the same conditions of physical environment (exposition of the slope, level of moisture and illumination, etc.). The same species of trees should be used as test trees when conducting census of lichens at different sites.

Censuses should be conducted by the method everywhere, which is with a measuring tape at the height of 150 cm.

It is recommended to determine lichen species while census taking. If it is not possible, then at least the number of different species found at the given site should be estimated. It is necessary to distinguish lichen species when carrying out a census at a single test tree.

Indexes of air pollution: the index of poleotolerance (IP) (only if species have been determined) and index of atmospheric quality (IAQ) should be calculated on the basis of field measurements for all studied sites. Then it is necessary to analyze differences among sites.

If both indexes of air pollution (IP and IAQ) have been calculated, then the results obtained by two different methods should be compared, as to whether these indexes reflect differences among sites and if data on concentrations of SO_2 in ambient air coincide at different sites.

Assessment of the vital state of a forest based on pine-tree analysis

This manual contains a procedure for environmental assessment of

forests based on the analysis of the morphological state of Scotch pine trees (Pinus sylvestris L). Key characteristics chosen as the main integral criteria are: the degree of crown defoliation, needle discoloration, number of cones and growth of end shoots.

Introduction

In the last two decades, mass damage and degradation of forests has been recorded in many European countries. Forest degradation is mainly caused by atmospheric pollution. Acid rains, high concentrations of sulphur oxides and nitrogen oxides as well as ozone in the air cause direct damage to plants; they lead to the worsening of state of forests.

The third session of the Executive Committee of the Convention on distant transboundary transfer of air pollutants under the aegis of the United Nations Environmental Programme (UNEP) approved and launched the **Programme of International Cooperation on research and monitoring of air pollutant impact on forests** (1985). The Programme is part of the Global System of

Environmental Monitoring. The program is based on the collection of comparable data about the state of forests at a national level and the subsequent exchange of this data, aimed at a better understanding of the problem.

The given procedure is relatively simple and reliable to use so it can be successfully applied in the practical environmental education of students. It is based on studies conducted on constant sites and can be used for both long-term monitoring and individual research.

The present procedure of study, monitoring and assessment of the vital state of forests is based on the **bioindication method**. The method consists of an assessment of the environmental status (general vital state of the forest) according to different characteristics of the studied living organism (in our case, pine). In other words, the indicator-species informs us about unfavourable environmental conditions by its appearance: its vital state. What environmental factors cause a certain response of the tree to unfavourable conditions is a separate, rather difficult question, which is not touched upon when the given lesson is performed.

It is recommended to use **Scotch pine** (Pinus sylvestris L.) as a main bioindicating species in the course of this research. If this species is not found in your area, you can substitute another species of pine.

Pine trees are one of the best model species to serve as bioindicators. First of all, the pine is very sensitive to even slight changes of environmental conditions, including environmental pollution. Second, pine is widespread in many forested areas of Europe, Asia, and North America; thus it is easy to find suitable study sites. It also simplifies the problem of comparability of data collected in different regions. Pine is also a convenient object for study by students due to the fact that pine is an evergreen tree and produces only one shoot a year, which makes observations much easier. From a methodical point of view, pine is a well-studied tree species.

The given task is divided into **three stages**:
1. Choice of sites and selection of trees for study.
2. Description of the general vital state (GVS) of trees.
3. Evaluation and interpretation of collected data, presentation of the study results.

The following items **are required** for the research: a compass, a measuring tape (one for each team of students), field glasses and description forms (one per group of 2-3 students).

Choice of a site and selection of trees for conducting measurements

As a rule, all long-term studies, especially monitoring studies, are carried out at **constant sites**. However, if a study takes place only once, it should be conducted at specific fixed natural objects; their choice should be as random as possible. Thus, we reduce the factor of a researcher's arbitrariness and create the conditions for external control and assessment for accuracy and reliability of collected data.

Choice of site location

A site for study of the vital state of trees should be situated in a relatively large forest massif; its total area should be **not less than 1 hectare** (100m x 100m). The site should be located in the heart of the forest and should not border the edges of the forest, forest roads or paths. It is recommended that a site is located at a distance not less than 25 meters away from the forest edges, roads or paths.

Selection of trees

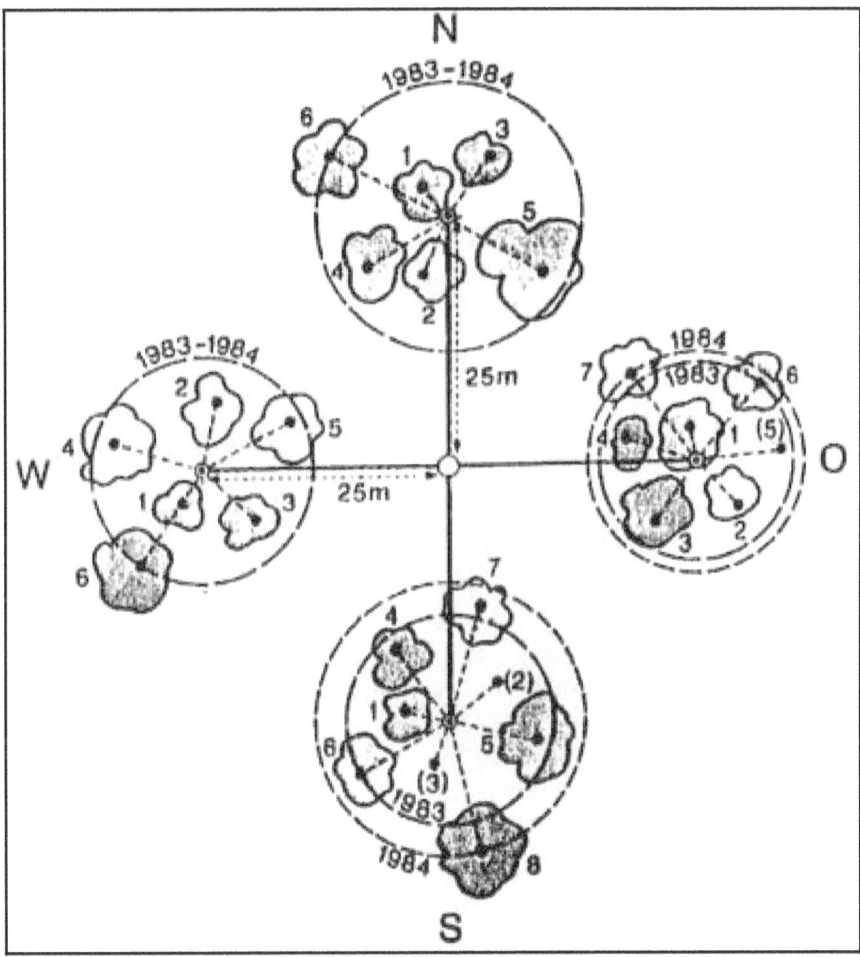

It is necessary to make provisions for a random sample of trees, independent from the researcher, while selecting trees for study. A system of random samples in scientific research, when chosen subjects (in our case, trees) represent the general state of the whole studied object (in our case, a forest), is called choosing a *representative sample*. The following system of tree selection is recommended for forming a representative sample (fig.1).

On his or her own, a researcher chooses only **the center of the site**. In order to make the search for site location easier, it is recommended to pick out a tree standing in the middle of a forest massif chosen for the location of the site, which can be easily found later. It is recommended to mark the tree with paint and to write a number or a site name on the tree.

A distance of 25 meters is measured from the center point (the tree) to the north, south, west and east with the help of a compass and a measuring tape. These points are marked with pegs, which are

hammered into the ground and painted. The height of the pegs should not be less than 20-25cm above the surface. Thus **corner points** will be marked on the site.

At the next stage of site layout it is necessary to pick out the six nearest trees around each corner point. These trees are given numbers and marked with paint or colour rings indicating direction to the corner point and the number of the tree. It is recommended to give trees letter-and-number indexes, where the letter stands for direction from the central point of the site, and a number indicates the number of the tree at this corner point. For instance, index N1 corresponds to tree #1 situated in the northern part of the site, index S6 corresponds to tree # 6 in the southern corner of the site ("W" stands for West, "E" for East).

Trees of the upper layers (1 and 2) form mature and immature stand. Underbrush and under-wood are not included in the description. In the mixed forests, only trees of the chosen species should be described. So, not less than six trees of the first layer should be described at each corner of the site (totalling up to 24 trees on the site, i.e. six trees at each of the four sides).

The following **conditions** should be mentioned; they are **significant** for your research:

- **If** there are so few trees at the chosen site for study that it is impossible to pick out **six trees** of the first layer within the distance of up to 12 meters from the center of the site, then it is necessary **to move the center** of the site to a denser part of the forest.

- **If** there are trees with easily recognizable signs of **mechanical damage** (breakage from snow or windfalls, large wounds on the bark) on the site, such trees should be **excluded** from a sample.

- **If** a **dry tree** has been found on the site while marking trees and it doesn't have clear signs of mechanical damage, such trees should be **included** in the sample as "old stand."

- **If** in the course of the study, some of the trees are exposed to mechanical damage or are cut down, then it is necessary to pick out the next tree closest to the corner point. It should be given the next consecutive number, and this replacement of the tree should be recorded in the field diary for the site. For example, if tree # 1 has been wind fallen at the site where there are six trees, it should be replaced with tree # 7, and not given the "old" number - # 1. This is necessary so that we do not mix up data about the "old" tree on the site (# 1) and the "new" one (#7).

- **If** some trees **die off and dry up** without any evident reasons, mechanical damage or snow damage, these trees **should remain** in the sample as "dead standing wood."

Period of description

As the procedure can be used for different tree species, the optimal period of study will be different for each of the various groups. The period when a description of the general vital state of pine trees can be carried out is practically unlimited. The period **from the end of August until December** is considered the best time for monitoring. Pines can be described once a year.

However, if you choose another tree species due to lack of pine in your area, then you must remember while interpreting the given procedure on your own, that spruce and fir trees produce **two shoots a year**, so it is necessary to conduct studies twice a year: in spring (May through beginning of June) and in autumn (October-December). Studies of **all defoliated** deciduous and coniferous tree species (for example larch) should be conducted at the end of the growing season, before leaves start to turn yellow. So, it is necessary to direct attention to the phase of growing season, but not to exact calendar date. **Small-leafed** wood species (aspen, birch, alder) **are not used** as bioindicators of environmental pollution.

Drawing a site passport

It is necessary to draw up a **passport** (indication card) for each site. The passport should contain the following information:

1 – geographical and administrative location of the site and a map of the area

2 – description of the site according to the following plan:

a) altitude of the site above sea level;

b) designation as a plain or a slope; if it is a slope, then its angle of incline and exposition should be recorded;

c) soil type.

3 – main data on the forest type at the site, including: tree species, crown density, average age of trees, under-wood and underbrush, grass vegetation.

4 – main data on chosen trees:

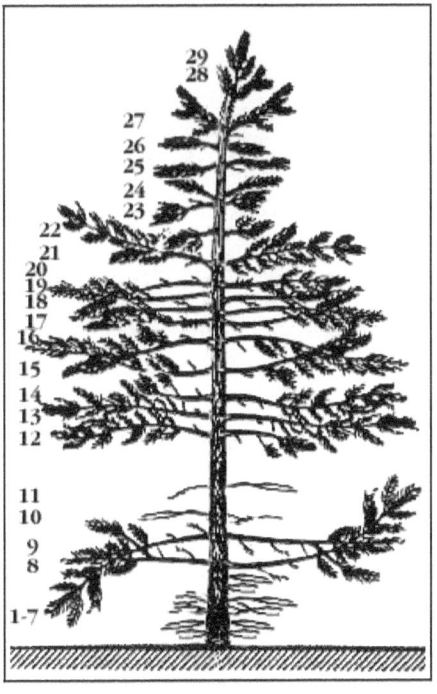

a) height and diameter of the tree measured at the chest level;

b) average age;

c) damage that the tree has sustained (mechanical damages and those caused by fungi and insects).

Age of trees can be estimated according to two methods: by counting year-rings (it is necessary to find a cut tree or to saw up a fallen tree) or by counting branch clusters. The second method is suitable for age estimation of pine trees; it is simple and reliable (for trees not older than 50-60 years). Each branch cluster on the tree trunk corresponds with a year's growth. In order to estimate age of a pine tree (with an accuracy of within two years), it is necessary to count precisely (with the help of field glasses) the number of branch clusters on the trunk and to add approximately 5-6 years.

Additionally, it is recommended to obtain the following data when conducting integrated studies and when teaching how to solve complex environmental tasks and perform monitoring:

About trees: average age of needles, maximum age of needles, average length of the top (terminal) shoot. Pine needles usually live for four years but in some regions, it is possible to find five- and six-year-old needles. Under conditions of strong environmental pollution, needles die at the age of two years. Average length of the top end shoot in favourable conditions varies within the range from 6 up to 15 cm depending on the region where the tree grows.

Anthropogenic factors: air quality (based on data of the nearest observation station), degree of soil compactness (caused by trampling), presence of clumps, and local sources of air and soil pollution (especially by sulphur and nitrogen oxides), acidity of precipitation.

Description of the general vital state (GVS) of trees

Before listing the GVS description rules, it is necessary to remember some biological features of trees and their use in bioindication. Pine trees (as well as most other tree species) have recognizable and measurable responses to worsening environmental conditions:

1) **Defoliation**, or the falling of needles (leaves), which can be visually evaluated according to reduction of usual crown density.

2) **Loss of natural coloration** (yellowing) of the crown.

These phenomena increase along with a worsening of the vital state of the tree until the tree dies off. Similar measurable responses of living organisms to various environmental changes lay the foundations of the bioindication method. The procedure of GVS assessment of trees is also based on the above-stated principles.

It is essential to find a point where **the whole tree** is easily observed in order to describe it correctly. When choosing a point of observation, one should remember that on a flat surface, the tree is better seen at the distance of one height of the tree from the trunk, whereas on slopes (in mountains or hills), a better view of the tree is found from a position corresponding with the middle of the crown.

Description of the vital state of trees consists in completion of the **special form** (see last page of the manual).

Filling in the description form for the vital state of trees

The description form of vital state of trees consists of two parts.

The first part is located at the top and provides **basic information** about the site. It contains very important information: number of the site, date of description, and information about the authors of the description as well as information about location of the site. This part of the form is important, because without it, all description data will make no sense, because they won't relate to a certain site, date and area of study.

The location of the site is its position with regard to geographical objects (towns, rivers, bridges, etc.) and local landmarks (boulders, detached trees, power transmission lines, etc.) When picking out local landmarks, it is necessary to take into consideration that they should serve as reference marks for several years, so it is recommended to choose such landmarks that are guaranteed to remain untouched for a long time.

A geobotanical name of the plant association, where all measurements are taken, should be written down in this column as well.

The main part of the form is **a table**, which contains different columns for the description of tree features indicating their vital state.

Column 1. "**Number of tree**" (Tree ID #) is filled according to marks on your site; the numbers of described trees are entered there (N1, N2…N6).

Column 2. "**Class of defoliation**" – this characteristic of the tree is evaluated visually with the help of field glasses. In order to determine the class of defoliation it is recommended to examine branches in the middle part of the crown.

Defoliation (crown density) is evaluated according to four major classes, where a certain percentage needle loss (or a degree of crown bareness) corresponds with an appropriate class:

Class 0 – defoliation is less than 10% (crown density is 90-100% of a normal tree)

Class 1 – light defoliation – 10-25% (density is 75-90%)

Class 2 – average defoliation – 25-60 % (density is 40-75%)

Class 3 – strong defoliation - > 60 (crown density is <40%)

Another method of determining the class of defoliation, which is

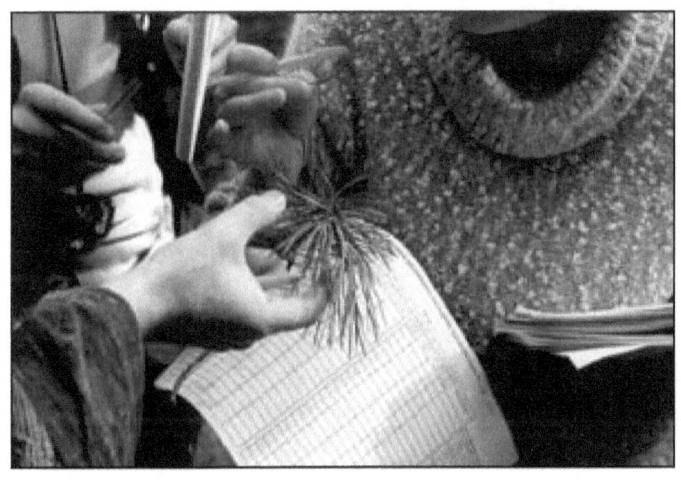

easier but not always effective, consists of an estimation of the **age of needles** that stay on their branches. It is known that pines produce one shoot a year, and its needles live for four years under "normal conditions" (at least in the central part of European Russia).

If four end shoots on one branch in the middle part of pine crown are covered with needles, then class of defoliation of the given tree is considered to be 0 (normal). If only three end shoots are covered with needles, then the class of defoliation is 1, if only two shoots are covered with needles, then the class of defoliation is 2, if only one end shoot is covered with needles, then the class of defoliation is 3.

Borders between shoots of different years can be observed visually with the help of field glasses according to branch clusters or an an gle between two shoots.

Column 3. "**Class of yellowing (discoloration)**". The degree of loss of coloration or "yellowing" of the crown is also evaluated visually according to four classes. A researcher tries to compare the observed colour of the given tree crown with a normal one as he or she remembers it or according to colours recorded in other descriptions.

Loss of natural colouring is estimated as percentage according to the following scale:

0 – no yellowing (loss of general colouring of the crown is 0-10 %0
1 – light (loss of 10-25 % colouring)
2 – average (25-60 %)
3 - strong (more than 60 %).

It is recommended to follow the procedure of determination of discoloration, which is described below. For the crown as a whole, first estimate the percentage of yellow needles without field glasses, then examine the quantity of yellow needles on branches with the help of field glasses. This recommendation is given due to the fact that color can be perceived differently in different weather conditions.

Column 4. "**New Cones**" - a number of new (unopened) cones on the tree is estimated visually according to the following scale and written in the mentioned column.

The number of cones is also estimated according to four-point scale: 0 – there are very many cones to 3 – there are no cones.

New cones are triangular cones. It is recommended to evaluate the number of these new cones on the tree with the help of field glasses.

Column 5. "**Old cones**". The number of old (opened) cones are estimated according to the same scale as new ones. Old cones resemble "hedgehogs" and counting them is easier to carry out with the help of field glasses as well.

Column 6. "**Growth of shoots**". Data of average growth in your area is required for an estimation of growth of top (terminal) shoots.

For example, the average annual growth of Scotch pine is 10-15 cm in the area around the Field Study Center "Ecosystem" in the region of Moscow. Small growth is considered to be less than 5 cm a year.

In our case, estimation of average growth is carried out according to a four point scale with an interval of 5 cm: 0 – growth is more than 15 cm, 1 – growth is 10-15 cm, 2 – growth is about 5-10 cm, 3 – growth is less than 5 cm. If there is another scale of average growth in your area, it is recommended to elaborate your own scale for estimation of average growth. The scale should also be a four-point scale, but will have another interval.

Column 7. "**Total points**". The algebraic sum of all points in columns 2 to 7 is calculated and entered here.

Column 8. "**General vital state**". This column is filled in the laboratory and refers to the processing of obtained data. How to fill this column is described below.

Column 9. "**Notes**". Additional information about your trees should be written in this column. The information can be useful in the course of the following interpretation of results; it includes, for instance, one-

sidedness of the crown (resembling a flag), the presence of a "stag-headed" tree, hollows, etc. Working information, for instance, damages to paint on the tree can also be recorded in this column.

Due to the fact that the evaluation of degree and type of defoliation, as well as the form of crowns is done visually, we recommend taking pictures of studied trees, to draw them and to make your own

schemes. You can make your own photo album of trees showing different types of needle damage (according to your visual assessment).

Some **important rules of conducting field descriptions** should be pointed out before going on with the explanation of data processing methods and rules of data interpretation.

Rule #1: Observers should be trained and get all necessary instruction; they must study the procedure thoroughly. Field glasses should be used for field descriptions.

Rule #2: No fewer than three observers should conduct the description. If their estimations differ, observers exchange their observation points and reach a mutual agreement in the course of discussion.

Rule #3: Studies are carried out in daytime with good illumination. Mistakes are possible at dawn and sunset, especially when determining the class of needle yellowing, due to patches of sunlight. Discoloration (yellowing) of needles is less noticeable in cloudy or rainy weather; in such weather crowns of trees look denser.

Data evaluation and presentation of study results

Data evaluation

Evaluation of data consists of the estimation of a value, which allows comparison and interpretation of the results of description. Evaluation can be carried out by two methods – by calculating the sum of points and according to GVS class.

Evaluation according to sum of points (Column 7): Calculating the sum of points is the simplest way of processing data. All values from columns from 2 to 6 are added in order to calculate the sum of points. Thus, the maximum value of points corresponds to a dying or dead tree. If the sum lies within the range from 0 to 5 points, the tree is healthy and rather viable. The scale is not precise due to unequal values in different columns; however, processing data and comparing it among different sites, using only the mean values of sums of points, is quite simple and easy. The smaller the average sum of points for all trees at the site, the better is the vital state of trees.

Evaluation of general vital state class (Column 8): Evaluation of general vital state based on the table and subsequent drawing of diagrams and their interpretation is a more flexible and correct method of data processing.

The general vital state of trees is determined on the basis of a combination of defoliation class and the class of needle yellowing using the following table:

Defoliation class	Class of yellowing		
	0 or 1	2	3
	General vital state class (GVS)		
0	0	1	2
1	1	2	2
2	2	3	3
3	3	3	3

We will find the class of GVS (from 0 to 3) **for each tree** at the site in the addition of "Defoliation class" taken from column 2 of the form and "Class of yellowing," taken from column 3 of the form.

Classes of GVS, which are equal to 4 and 5, correspond with a "new" and "old," or dead standing tree accordingly. Each class of GVS corresponds with a mean statistical value of the number of years until the tree dies off completely. Class zero (points) of GVS corresponds with a period of more than 20 years, class 1 corresponds with a period from 10 to 20 years, class 2, three to nine years, and class 3 corresponds with a period less than three years.

It should be mentioned that this is only a general statistical forecast of development of the tree under stable environmental conditions, so real terms can differ from proposed mean statistical values.

Visual presentation of obtained results

A method of drawing column diagrams is applied for the visual presentation of results following the interpretation of data. Diagrams can be drawn separately for defoliation and loss of natural coloration, however, in order to provide final evaluation of data and present results of conducted studies, it is necessary to analyze an integral index – GVS.

Let's list some **rules** in order to make drawing up of diagrams easier:

1) The occurrence of trees with different GVS values is plotted along the vertical axis. It is recommended to use a percentage scale. Before that, it is necessary to determine the percentage of described trees with different GVS values.

2) Columns of the diagram should be arranged in such a way, that you can illustrate the dynamics of the studied process. For instance, if it is a monitoring of time changes, then a year of studies increases from left to right, if it is a spatial analysis, then diagrams are arranged from left to right according to the distance to the source of potential pollution.

3) Classes of GVS are arranged upwards as GVS increases (indicating a worsening of the vital state). The higher the GVS value, the darker the columns are shaded.

4) If it is a colour diagram, then the colours of columns change from green to red according to increase in GVS class. Dead trees can be marked with black.

Procedure for drawing a column diagram

Using the following example, let us illustrate the procedure of drawing a column diagram. Suppose that 24 trees are described at the site (six trees on each of the four sides of the site). Among them four trees got 0 points of GVS (these trees are healthy), five trees received 1 point, four trees-2 points, four trees –3 points, four trees- 4 points ("new" dead standing trees) and three trees are characterised with the GVS class equal to 5 (old dead trees).

It is more convenient to fill the data into the following table:

GVS value	Number of trees at the site with the same GVS value	Percentage of trees with the same GVS value	Accumulated sum, %
0	4	16.7	16.7
1	5	20.8	37.5
2	4	16.7	54.2
3	4	16.7	70.9
4	4	16.7	87.6
5	3	12.4	100
Sum	24	100	

The Column "**Accumulated sum**" is required in order to simplify drawing a diagram; it contains figures indicating borders between areas on the diagram. They are obtained by adding the percentage value in the given line with the percentage values in all previous lines.

Boundaries of areas should be marked on the column diagram using the column "*accumulated sum*", whereas areas themselves are shown on the diagram with the help of colouring, marks, shading, etc.

As a result, we end up with the following diagram (Picture 1):

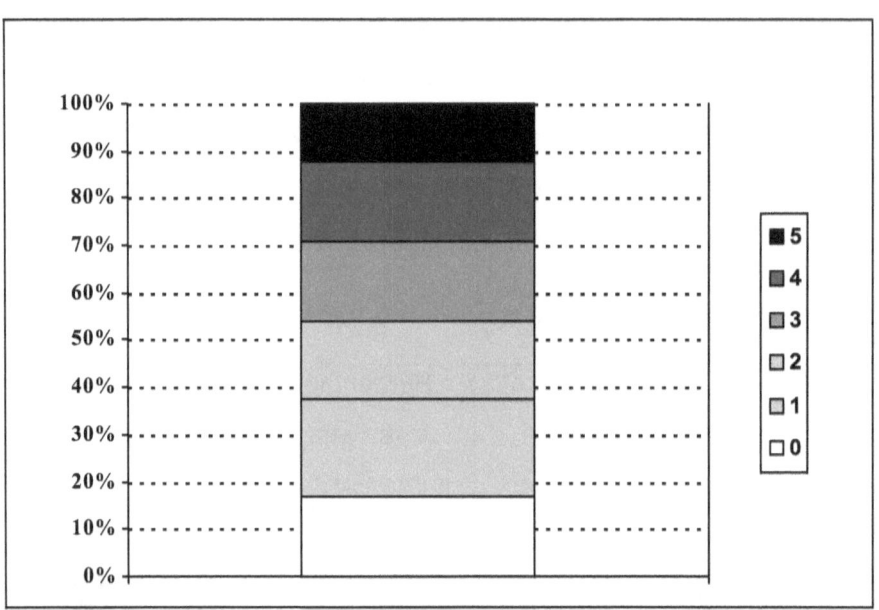

Picture 1. An example of a column diagram of the general vital state of the forest at one site.

General scheme for organization of studies

As the **goal** of this lesson is not only familiarization with a procedure, but concluding real studies, then at least two sites should be laid down and described within the framework of the present educational task. In order to make a comparison that is interesting and real, it is necessary to choose two sites in the forest, which differ greatly according to their level of anthropogenic load: for instance, sites located at different distances away from an obvious source of pollution (human settlement, a factory, highway, etc.). It is recommended that these sites be situated along one line in the direction of prevalent winds.

If there are no evident sources of air pollution, then it is recommended to choose sites for comparison that are used differently in forest service (thinned forested areas and littered sites) or in agriculture (with cattle grazing and without it), sites with different levels of anthropogenic use (for example, trampling) and so on. In any case, sites with approximately **similar** types of vegetation and trees of the same age should be selected for comparison. An ideal variant is the choice of sites located in two different pine tree forests of the same age.

Analysis and interpretation of the research results

Below we show as an example diagrams drawn on the basis of measurement data obtained at two sites in the vicinity of the Field Center "Ecosystem" in the region of Moscow. Diagrams are drawn and arranged so that they provide an opportunity to determine the influence of only one changeable parameter. Picture 2-a shows differences in the state of the forest, depending on the distance away from the human settlement.

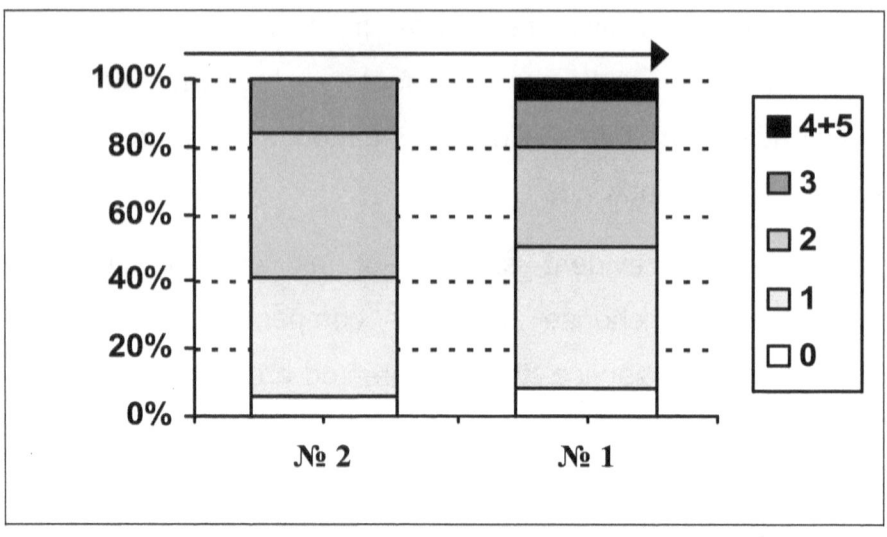

closer to the settlement →→→ farther from the settlement.

Picture 2-a: Dependence of forest state on distance away from the settlement.

The picture shows that the general state of trees at the site closest to the settlement (site #1) is worse than general state of trees at the remote site (21% of trees with GVS class 3-5 at site N1 correspond with only 16% of trees at site N2). At the same time, the number of healthy trees at this site is also greater (50, 5% of trees of GVS class 0-1 compared to 41%).

Taking into account the fact that the number of "stag-headed" trees is higher at site #1, we can make a conclusion about increased anthropogenic impact on the nearest forest massif in comparison with the remote forested area.

It cannot be determined exactly what anthropogenic factor has prior significance here with the help of this described method. We can only suppose, that over-compression of soil caused by trampling plays the leading role. A diagram drawn on the basis of similar description data, collected for several years of observation, is shown in picture 2-b in order to illustrate the dynamics of time changes. Descriptions of two sites are generalized for each year.

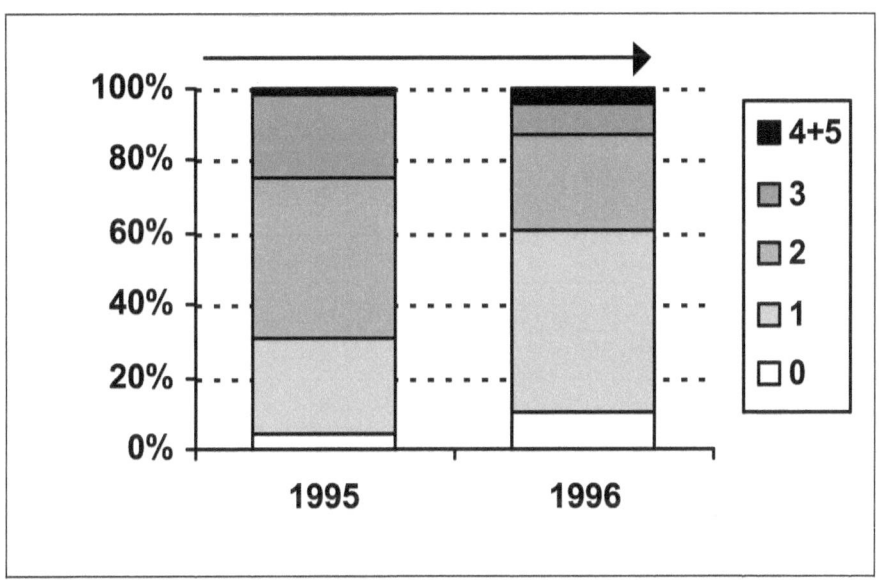

Picture 2-b Year-to-year changes of the state of pine trees

Diagrams for two years show that a share of healthy trees increases over the course of time. At the same time, the number of ill trees also increases. Only a share of "average" trees decreases.

Apparently, it is evidence of the strengthening of mutual intraspecific competition. In our example, sites are located in pine tree stands that are 40 years old. While trees are growing, space competition intensifies and trees begin to depress each other. Stronger trees improve their vital space at the expense of deaths of weaker trees.

56

The Description of Forest Health Based on Individual Tree Data

Site № _____ Date _____ Students _____

Location _____

1	2	3	4	5	6	7	8	9
Tree ID #	Class of crown defoliation (0-3)	Class of discoloration (0-3)	New cones (0-3)	Old cones (0-3)	Growth of shoots (0-3)	Total points (0-15)	General vital state (0-5)	Notes

Note: There should be 24 lines in the table, which correspond to the number of described trees on the site.

<u>Suggestions for filling in the form</u>:

Class of crown defoliation: 0 – normal (only 10 % of needles fell down, four-year-old needles are found); 1 – light (10-25 %; three-year-old needles); 2 – average (25-60%, two-year-old needles);

Class of needle yellowing: 0 – normal (0-10 % of needles), 1 – light (10-25 %), 2 – average (25-60 %), 3 – strong (>60 %).

Number of cones (new and old): 0 – very many, 1 – many, 2 – a few, 0- none.

Growth of shoots: 0 – very large (15 cm), 1 – large (10-15 cm), 2 – average (5-10 cm), 3 – small/no (5 cm).

The table for evaluation of general vital state (GVS):

Defoliation class	Discoloration class		
	0-1	2	3
0	0	1	2
1	1	2	2
2	2	3	3
3	3	3	3

Assessment of ecological features of meadows on the base of vegetation cover

This manual contains a procedure for bioindication of ecological  conditions in meadow plant associations according to the species composition and abundance of plants. Without any geographical, physical, chemical or historical studies, it is also possible to determine important characteristics of habitat physical conditions.

Introduction

This research is one of the most difficult and laborious studies within the given series, as it requires **knowledge** of herbaceous plant species, **skills** to make geobotanical descriptions of meadow plant associations, as well as the **ability to process** quantitatively obtained data.

The present lesson is based on one of the most interesting bioindication methods. **Bioindication** is the determination of different environmental characteristics according to living objects. In the given case, objects of bioindication are meadow plant associations. This procedure allows us to determine the main **geographical and physical characteristics** of these plant associations on the basis of only the description of the **species composition** of plants growing in the meadow, without conducting any geographical (geomorphological, geological and soil), physical, chemical or historical studies.

Long-term work of the Russian soil agronomist and botanist L.G. Ramensky, who carried out large-scale studies in 1940-50s, is used as the basis of the procedure. Ramensky analyzed vast materials collected in different regions of the country where he studied the species composition of meadows and their physical properties at the same time. Comparing species composition and abundance of plants in meadows located in areas with different physical properties, he determined colonization patterns of different meadow types by different plant species. Due to his works, the "Comparative ecological tables of plants" were developed, which are described in this manual. Using the tables, a researcher can describe values of certain physical properties of the area under study on the basis of plants found in the meadow.

L.G. Ramensky studied **five ecological characteristics** ("scales") of meadow plant associations: 1) scale of "moisture" (denoted by a letter "M" in the tables); 2) scale of moisture variability (MV); 3) scale of "active soil richness and salinity" (RS); 4) scale of alluvialness (A) and 5) scale of "pasture digression" (PD). What the scales are, what they characterize and how researcher can evaluate the position of a given meadow plant association on the basis of each of the scales – all these issues will be discussed below in the section called "Characteristics of scales-gradations of ecological conditions."

Students are required to have the following equipment for conducting the studies: forms for geobotanical descriptions (or field diaries or notebooks), complete (scientific) field guides of herbaceous plants and comparative ecological tables, given in the present manual.

General organizational plan of studies

It is recommended to organize the students conducting these studies with the example of 3-5 meadow plant associations that are known to differ in physical properties of their environments. For instance, it is possible to include and describe meadows found at different distances and at different heights according to the riverbed. This will

provide an opportunity to compare meadows with different types of moisture, moisture variability, soil richness and alluvialness.

The total amount of work, i.e. the number of different types of meadows studied in the course of research depends as usual on the availability and qualification of "labor resources" and time. Two or three students can carry out a standard geobotanical description of a meadow. Each team of students conducts the description of two or three types of meadows found in the vicinity of the field studies center. When planning the studies, it is necessary to take into account the fact that fieldwork takes relatively little time (not more than an hour for description of one meadow); most of your time will be spent processing the field material (determination of unknown species and work with tables).

Requirements for the fieldwork in the course of given studies include students' knowledge of all the species of **herbaceous** plants or the availability of complete (scientific) field guides as well as the skills to conduct description of herbaceous vegetation.

Technique for geobotanical descriptions of meadow vegetation

Description of meadow vegetation is carried out according to a standard scheme.

A **site for description** of 1 m x 1 m in size is laid down in an average, middle part of the meadow. The corners of the site are marked with pegs for convenience and visibility, and the borders of the site are marked with strings (or one meter long sticks).

Description of the herbaceous layer is carried out within the boundaries of the site. The description of the herbaceous-shrub layer includes compiling a list of plant species at the given site with rough estimations of their abundance.

A simple indicator of herbaceous plant abundance is the measurement of **projective cover**. Projective (or descriptive) cover for herbaceous plants is actually the same parameter as crown density for arboreal and bush layers.

Projective cover is expressed as a percentage and is determined for each species separately. Thus the sum of projective cover values of all the species can be more than 100 if leaves are overlapping. The total projective cover can be less than one hundred percent if some soil surface is left uncovered with plants. This situation is more typical for the early spring season. The expected accuracy of projective cover estimations should be 5 %.

Outwardly, the **geobotanical description** of a meadow looks like a list of plant species with an indication of values of projective cover as a percentage next to each plant species. A standard "form of geobotanical description," which can be found in any of the above-listed botanical manuals, can be used to keep records. As the form was first developed for the description of forest vegetation, thus the description of meadow plant associations can be done simply in a field diary or in a notebook. It is important that beside the geobotanical description itself, each meadow plant association is given a "general characteristic" date and place of description (administrative and geographical location), description of surrounding area (what biotopes are found near the meadow and what plant associations the meadow adjoins), and the surname of the authors of the description.

Unknown plant species encountered during the description are sampled for the herbarium and taken back to the field studies center or classroom for further determination. Thus they should be given a certain number (index) in the description form, which is replaced with a species name after determination.

When the team of students comes back to the field studies center, they process the field materials. First of all, they determine all the species that were registered within the site. Then the students copy out the "limiting levels" for each plant species with its projective coverage according to the table provided below. They estimate an average value of the limiting level for the given meadow plant association. Such procedure is carried out separately for each of the scales of ecological conditions. Indexes found for different meadows are then compared.

All data required for processing the description records is given in the following separate sections.

Description of scales of ecological conditions

As it has been said above, L.G. Ramensky studied **five characteristics** of ecological conditions in meadows, which he called "scales": *moisture, moisture variability, soil richness and salinity, alluvialness and pasture digression*. Ramensky divided each of the scales into gradations – **from 6 up to 12 gradations** in each scale. Each gradation is provided with a description of typical ecological conditions of the meadow: its location within the relief, type of plant association, general appearance, species composition of prevailing (background) plant species, soil type, soil acidity (pH), mean annual precipitation, regularity of flooding with river waters, intensity of anthropogenic load, etc.

This section provides general descriptions of chosen **ecological characteristics** of meadows as well as **descriptions of meadows** according to each gradation of Ramensky's scales.

Scale of Moisture (M)

This scale characterizes the level of **total** moisture in the habitat. It is not momentary moisture of soil or air, but a complex, integral characteristic, which includes total average annual precipitation, moisture level of soil and plant association type. Ramensky distinguished 12 gradations within the given scale and each gradation corresponded with a specific soil type and plant association.

Degrees 1-17 reflect *desert* moisture, conditions of extreme deficiency of soil moisture with very sparse vegetation, quite scanty in species composition (*sagebrush* species, annual and perennial *shrubby saltworts, ephemerals* and other plant species). Such soil types are typical for these conditions: *takyrs*, desert light *sierozems*, light-brown and brown soils. About 150mm of precipitation per year and less is observed in the extremely dry climate of desert and semi-desert zone (along *solonetz soils, dry slopes*, etc.). There is almost a constant lack of moisture in soil. Dry, non-irrigable farming is almost impossible.

Degrees 18-30 show *semi-desert (desert-steppe)* moisture close to the desert type. Annual precipitation total is 150-250mm. Soils are light-chestnut, brown, light-brown, *sierozems*; dominating species are *sagebrush, xerophilous grasses (fescue, wheatgrasses, feather grasses)*. These conditions of the semi-desert zone are found in the zone of dry steppe along slopes and *solonetz*, whereas in the desert zone they are in more moistened areas. Dry (non-irrigable) farming is very unstable has very little productivity.

Degrees 31-39: *dry-steppe* moisture. Annual precipitation is 250-300mm, soils are dark-chestnut and southern black earth. Prevalence of small grass species is typical – *wheatgrass, feather grass*, and on some sites – *Austrian wormwood*. These conditions are also found in zones of semi-deserts and deserts along more moistened dry depressions (outskirts of firths, swallow holes), and in the zone of middle steppe – along southern slopes and solonetzs. This moisture is enough for dry farming on the condition that techniques of "dry farming" are applied (accumulation and preservation of moisture in soil, its economical use, cultivation of fast ripening crops, etc.)

Degrees 40-46 are the range of moisture of **middle steppe**, which corresponds with conditions of large feather grass steppes. Soils are common black earth and medium *chernozems*. Such habitats are found in more southern zones along depressed sites with accumulating waters, in the zone of meadow steppe (forest-steppe) – along southern slopes. Habitats of middle-steppe moisture are quite suitable and used for field-crop cultivation, but they are often exposed to drought.

Degrees 47-52 reflect moisture of *wet steppe or meadow-steppe*. Rich meadow steppes and steppe dry meadows, as well as variants of dry forests – pine and oak types (the zone of forest-steppe). Soils include chernozems: thick, fertile, degraded, gray forest-steppe loamy soils, black earth- meadow soils. They are found in more southern zones along depressions without excessive moisture – along high parts of river valleys, swallow holes, firths, and the most drained dry slopes. Farming, which is quite sustainable and provided with sufficient moisture, transpires in these habitats; there are good hayfields and pastures, which are, however, exposed to lack of moisture and in drier years get dried out or burnt.

Degrees 53-63: moisture of *dry and fresh meadows and forests*. Soils are *meadow (turf), podzolic, brown burozems*. It corresponds with drained habitats of the forest (forest – meadow) zone. They are found in more southern zones along depressions of different types – along river valleys, swallow holes and firths. In dry years, herbage also suffers from lack of moisture. Farming characterized by quite sufficient moisture is created in these habitats; however, it also suffers from lack of moisture in dry years.

Degrees 64-76: *damp meadow* moisture. Soils are usually without signs of **gleyification** or have low gleying. Prevalence of the best meadow grass species and clovers is observed at these meadows (upon the condition that there are rich meadow soils). Damp meadow habitats predominate on rather drained flat and elevated parts of valleys in forested zones. Habitats of damp meadow moisture are located in southern zones along depressed relief elements – along the bottoms of gullies, firths and river floodplains. There, high-yield grassland meadows are formed; sometimes they suffer from lack of moisture only in the second half of the year. They are the best habitats for meadow plant species. Cultivation of field crops provides high yield there, but they suffer from excessive moisture in some years, so some drying is required – facilitation of surface run-off. Under conditions of very wet or damp, evenly distributed moisture and sphagnum peat substratum in the forested zone, variants of drier highmoor bogs (with pine trees, heather, etc.) also belong to these degrees of moisture.

Degrees 77-88 show *wet meadow* moisture: very wet meadows and forests as well as relatively dry peat highmoor bogs. Soils are highly gley or peat. Good meadow grasses suffer from excessive moisture. Such areas require drying, which can be provided with facilitation of surface run-off in some cases.

Degrees 89-93: *bog-meadow* moisture. Peat-land meadows and forests, low water bearing bogs. Such moisture is found within the forested zone along non-drained plains, along lowlands and river floodplains, and in southern zones – along superfluous flooded firths or their central parts, along terrain (less often – central) and riverbed parts of floodplains.

Degrees 94-103: *bog* moisture, which are average- and highly water-bearing bogs. The following plant species are typical for such areas: marsh cinquefoil, buckbean, sedge, etc.

Degrees 104-109 are habitats of *plavni* (low parts of down-stream valleys of Russian rivers covered with reeds and trees) and *riparian-aquatic vegetation* (cattail, cicuta, and others).

Degrees 110-120: habitats of *aquatic vegetation* – water lilies, pondweed, etc.

Scale of moisture variability (MV)

This scale characterizes the constancy of water balance in the area, where the given meadow plant association is located. Six gradations of the scale are distinguished.

Degrees 1-4 – *highly provided* water supply. They cover areas with even moisture without any crises. It is fully realized in a humid climate, in areas that are fed with groundwater of a constant level that lies at low depths. The high water supply of meadow openings is favored by their surrounding forests, which reduce evaporation. We often find such habitats along edges of lowlands and river valleys. The habitats under review are characterized with an overgrowth of green and sphagnum mosses as well as a prevalence of typical *mesophytes* with thin wide leaves.

Degrees 5-6 – *average* water supply. Plants typical for habitats with highly provided water supply, combine with plants that are more resistant to temporary water deficit. Habitats of the type are widespread in the humid climate of the forested zone. In the case of a provision of water supply with close groundwater (near strings outlets), such conditions are created in less wet areas.

Degrees 7-8 – *variably provided* water supply. These conditions are even wider spread and they are the most typical for different habitats of the forested zone. Adequate provision of the areas with water varies distinctly in different years and vegetation periods, but the variability is not so high as to cause the formation of corresponding protective adaptations of plants. There are no tangible crises in supplying plants with moisture. These habitats found farther to the south are associated with northern slopes, and with outlets of groundwater; this association becomes stricter and narrower while moving deeper into the steppe zone.

Degrees 9-11 – *moderately variable* moisture. Such habitats include the best meadow grass species and legumes; moss cover is poorly developed in such areas, as variability of moisture is unfavorable for most mosses. Habitats of this type are typical not only in the forested zone but also in the forest-steppe and even farther to the south. Habitats of such moisture variability are favorable for the growth of plants, with a sufficient moisture level. Drying up and lowering of the groundwater level, which take place from time to time, promotes soil aeration, ameliorates nitrification and other microbiological processes in the soil.

Degrees 12-15 – *strongly variable* moisture. They include meadows located in river floodplains and other depressions of different types, which are flooded in spring and drained by the river in summer, as well as many continental habitats, especially in the steppe zone. High variability of moisture is promoted by the salinity of soils and presence of a distinct waterproof alluvial horizon in the soil. The plants that are mainly associated with flooded areas are dog-grass, meadow foxtail and others, are typical for such soils.

Degrees 16-20 – *sharply variable* moisture. Such conditions are crated within firths and flood lands of large rivers, which are flooded late and for a long time.

Scale of active soil richness and salinity (RS)

Active soil richness is the provision of soil with plant nutrition elements in a form that is mobile and easily assimilated by plants (soluble salts, bases absorbed by colloids, easily mineralized nitrogen compounds, etc.). There are ten gradations within the given scale.

Degrees 1-3 of the scale of soil richness and salinity: *extremely poor soils* and *peat* (*oligotrophic*). Soil reaction is an acid reaction: pH = 4.0-4.5, but in the case of peat it can be higher. Mineral soils are highly leached; they are quite often sandy soils. Habitats of the poorest dry meadows found in the forested zone (with *cat's foot* and *matgrass*), on sands in pine forests, highmoor bogs with sphagnum mosses and peat of low ash content (2-6%) belong to this group.

Degrees 4-6: *poor soils and peat*. Soil reaction is acid: pH=5.0-5.5, but in case of peat it can be higher. Mineral soils are leached; they are quite often sandy and loamy types of soil. The areas include habitats of poor dry meadows in the forested zone (with *matgrass*), *pine forests* and *subors* (a mixed forest on transitional, relatively poor soils), *highmoor* and transitional bogs with *sphagnum mosses* and peat of relatively higher ash content (6-9%).

Degrees 7-9: *rather poor soils (mesotrophic)*. Soil reaction is sub-acid (pH 5.5-6.5). Soils are usually *podzolic* (ashen-grey), *sod-podzolic, podzolic-gley*, *peat* and others. Dry meadows are located in the forested zone, poorer lowland meadows, and bogs with peat of increased ash-content (8-12%).

Degrees 10-13: *rather rich (fertile) soils*. Soil reaction varies from sub-acid to neutral (pH 6.0-7.5). Soils are *meadow, forest-steppe, loamy,* and *leached chernozem.* Habitats are located in the floodplain and lowland meadows, bogs, as well as in steppe and oak forests.

Degrees 14-16: *rich (fertile) soils*. Soil reaction is neutral (pH 7.0-7.5). Soils include *thick common and southern chernozems, non-saline, chestnut, brown and sierozems, alluvial-active meadow, lowland meadow* and other poorly leached soils, which are rather rich in nutrition elements and free from hazardous salts. Steppe, partly semi-desert and desert vegetation is found there. Vegetation of the best floodplain and lowland meadows, oak groves, and some bogs is associated with such habitats.

Degrees 17-19: *low-saline soils*. Soil reaction is alkaline (pH 7.5-8.3). They are widespread among meadow soils along river floodplains, along depressions; low salinity is also widespread along plains and depressions in steppe, semi-desert and desert zones on soils of corresponding types.

Degrees 20-21: *medium-saline soils*. Reaction is sub-alkaline (pH 7.5-8.3). They are more often meadow saline soils with a distinct content of **sulfuric** and **chlorous** salts in the upper half- meter layer.

Degrees 22-23: *highly-saline soils (salt marshes).* They are meadow salt marshes at high levels of moisture. Their soil reaction is usually alkaline (pH up to 9.1). The upper half-meter layer contains a high content of sulfur, chlorous and other salts.

Degrees 24-28: *sharply saline soils (salt marshes).* They are meadow salt marshes at high levels of moisture. The upper half-meter layer contains a high content of salts (several percentages).

Degrees 29-30: *persistent saline soils (persistent salt marshes, sors).* Accumulation of salts near the surface and on the soil surface reaches such an amount, that halophytic vegetation becomes very thinned out or it is completely lacking; soil surface is usually covered with a salt crust of different thickness.

Scale of alluvialness (A)

This scale characterizes the degree of "*alluvialness*" of a specific area, where this meadow plant association is found. Let us remember that alluvium is river deposits; thus, alluvialness is the susceptibility of the area to periodic flooding with river waters, or the consequences of such floods in the past. There are seven gradations in the scale.

Degree 1 belongs to *non-alluvial* habitats *without deposition of warp* or with slight traces of it.

Degrees 2-3: very *low-alluvial (or dealluvial)* habitats, where almost all plant species reside. Warp is about 1mm (degree 2) and up to 2-3mm (degree 3).

Degree 4: *low-alluvial* (warp of about 2-5mm). Certain selection becomes obvious, as "non-alluvial" species become suppressed.

Degrees 5-7: *moderately-alluvial* habitats. Warp deposit for the 5th degree is 0.5cm, for the 6th degree, 1cm and for the 7th, 2cm.

Degree 8: *highly-alluvial*. Deposit of warp is 2-4 cm. Alluvial plant groupings, with a prevalence of creeping- rhizome gramineous species are the most expressed.

Degree 9 characterizes *excessive deposition of warp* of 5-10cm thick. Vegetation cover is thinned out and annual weeds penetrate into it; their seeds are brought along with warp particles.

Degree 10: thickness of warp is *catastrophic*, usually 10-15cm and over. Aboriginal vegetation is suppressed and it hardly forces its way out; alien weeds prevail or there are few of them; the area is bare. Similar catastrophe causes even thinner warp, if it is deposited on the layer of fallen tree leaves.

Scale of pasture digression (PD)

This scale characterizes *the intensity of grass coverage destruction* by herbivorous animals. While developing the scale, L.G. Ramensky studied, first of all, the influence of cattle on grass cover. However, this scale also characterizes the natural destruction of the grass cover by wild herbivorous animals, as well as (but not as precisely) the trampling of grass cover by people. This scale is subdivided into seven gradations.

Degrees 1-2: impact of pasturing is *slight or completely lacking*. They are meadows where pasturing and sowing do not have noticeable impact. They are characterized with high herbage, where the crowning grass species of mowing type share predominance with large and broad-leaf motley grass species or they are even suppressed by them. Species composition of aboriginal motley grasses includes different species of *geraniums, Umbelliferae,*

Compositae, Rosales, Ranunculaceae (meadow-rue, globe-flower). It is an *initial stage* of pasture digression.

Degrees 3-4: *low impact* of pasturing, *mowing* stage. Pasturing as well as early mowing oppress motley grass, crowning grass species (*timothy, meadow fescue, Hungarian brome, slough grass* and to a lesser degree – *meadow foxtail*) take an advantage, i.e. species, which are the most needed at hayfields. It is a *mowing stage* of pasture digression.

Degree 5: *moderate impact* of pasturing. Native motley grass species fall out (with the exception of some plant species, which become adapted), pasture weeds appear and spread out; crowning mowing grass species begin to replace low pasture species. It is a *semi-pasturing* stage of pasture digression.

Degrees 6-7: *strong impact* of pasturing. It is characterized by the prevalence of lower pasture grass species (*meadow-grass, red fescue*, different species of *bents*) and *legumes* (*shamrock*); there are many perennial weeds (*dandelion, autumn hawkbit, silverweed, golden-cups* and others). *Pasturing stage.*

Degree 8: *half-destruction (half-trample)*, is close to the previous stage. Crowning grass species have fallen out to a certain extent; weed perennials spread out, replacing pasture grass. Herbage becomes thin; it includes annual weeds – *knotweed, annual meadow-grass, blindweed* and others. Thorny perennials become widespread – thistles.

Degree 9: *destruction (trample)*. The herbage is heavily thinned out; it is mainly composed of *knot-grass* and other trampling annual plants.

Degree 10: *absolute destruction (trample)*. The soil is bare; there are single different plants (weeds), which cover an insignificant part of the area.

Procedure for using ecological scales

The scales proposed by Ramensky and the tables made on the basis of them **characterize environmental conditions** of plants growing in natural herbage for each of the plant species found. Using the tables, students can determine habitat conditions **on the basis of vegetation**.

The tables are alphabetical lists of plant species, where the ecological range of the species habitat (in terms of gradations of available ecological scales) is described for each species depending on projective abundance of the given plant within the specific plant association.

There are several procedures for processing descriptions in order to determine characteristics of the habitat on the basis of geobotanical descriptions of vegetation (where projective abundance of each plant species within the site of one square meter is given). However, different table modifications of the *limitation method* are used more often.

Let us review the procedure on the basis of a specific plant association.

First, we write out all plant species found within the site in decreasing order of their abundance (projective cover - PC):

Species	PC
Red clover	54%
Yarrow	34%
Koeleria Deliavina	29.5%
Meadow-grass	7%
Sheep's fescue	6%
Lady's bedstraw	3.1%
Vernal sedge	2.7%
Spiraea	1.1%
Upland crane's bill	0.5%
Field scabiosa	0.5%
Salsify	0.4%

For the convenience of the following use of comparative ecological tables, let us give conventional letter symbols to the plant species in accordance with their projective abundance. This letter scale of projective abundance is a standard scale used by botanists (Scale of abundance by O. Drude). This scale is given in the heading of Table 1 (see the link below).

Thus, our description will look as follows:

Species	PC
Red clover	m
Yarrow	m
Koeleria Deliavina	m
Meadow-grass	c
Sheep's fescue	c
Lady's bedstraw	c

Vernal sedge	c
Spiraea	n
Upland crane's bill	n
Field scabiosa	n
Salsify	n

Now, with the help of **ecological scales** we will find limiting levels for each of the listed plant species in the column of corresponding projective abundance (Table 1, can be downloaded from http://ecosystema.ru/04materials/manuals/ramensky.doc). Species are listed in the table in alphabetical order by their Latin names. For instance, in scale "M" (moisture) limiting levels of moisture for meadow-grass will the values "60" and "89", as its abundance corresponds with "C" class (abundant) in compliance with Ramensky. In the case of higher abundance (m), limiting levels for the species will be "63" and "68" and at moderate abundance (n) – "57" and "92". On the whole, the limiting degrees for the given plant association look like the following (scale "M"):

Species	Limiting levels	
	From	To
Red clover	58	72
Yarrow	58	63
Koeleria Deliavina	50	56
Meadow-grass	60	89
Sheep's fescue	17	59
Lady's bedstraw	45	65
Vernal sedge	49	73
Spiraea	45	64
Upland crane's bill	55	89
Field scabiosa	42	64
Salsify	47	59

In order to determine **limiting levels**, which are located at the turn of each other or overlapping each other, it is necessary to arrange them from the side of the dry wing of the row ("from") in decreasing order of changes whereas limiting levels from the side of the wet wing of the row ("to") – in increasing order. As a result, we will obtain two rows of successive levels, where the wettest limitation levels of the dry wing of the row are confronted with the driest limitation levels on the wet wing of the row in pairs:

from	60	58	58	55	50	49	47	45	45	42	17
to	56	59	59	63	64	64	65	72	73	89	89

The intersection of these two rows of limiting levels takes place at degrees 58 and 59. This means that the degree of moisture is determined as 58.5. The pair of figures to the left (60 and 56) provides the value (58) closest to the solution. The solution should be checked at 4-5 pairs of figures and then the average value should be estimated: such checking also results in the value of 58.5.

When the **ecological degree of the plant association** under study is determined according to this procedure, then you can limit yourself to an estimation of the average of 4-5 pairs of figures, which are located to the left and the right of the intercrossing pairs. The records can be kept in the following way:

From	To	Calculations	Result
60	56	116:2 =	58
58	59	117:2 =	58.5
58	59	117:2 =	58.5

55	62	118:2 =	59
50	64	114:2 =	57
	Average result - 58.5		

If the average of the left or the right pair of figures (among the pairs which adjoin the Intersection) deviates greatly from other figures, then it is excluded from the calculation of the average value. This variant of the procedure is less precise, but it allows us to avoid gross errors.

If estimation goes correctly and all the data are in accord with each other, then there is no need to write out the left column of decreasing figures and the right column of increasing values completely in the course of the practical work. It is enough to write out 5-6 pairs of the top figures: the figures that allow us to find the solution.

The desired solution according the scale of "M" is 58.5; it indicates that the specific plant association can exist under conditions of dry and fresh meadows on corresponding soils.

The **position of the plant associations** according to other scales ("MV", "PD", "A", "RS") is found similarly.

It is possible to give a more complete verbal description of the studied meadow plant association according to the above-described scales (Section "Description of scales of ecological conditions").

Assessment of environmental state of the forest based on leaves' asymmetry

This manual contains procedure for assessment of environmental state of the area, in particular, forests, based on integral characteristics of asymmetry of tree leaves. The procedure is based on the theory that difference between the left and the right halves of a leaf correlates with the degree of total environmental disturbance.

Introduction

The procedure used in the course of present research is based on the theory of "**development stability**" ("morphogenetic homeostasis"), elaborated by Russian scientists A.V. Yablokov, V.M. Zakharov and others in the course of studies of nuclear pollution effects after the Chernobyl disaster. Those scientists proved that stress-impact of different types caused changes in development *homeostasis* (stability) of living organisms and those changes could be evaluated on the basis of disturbance of *morphogenetic processes*.

Main indicators of changes in homeostasis of morphogenetic processes are indicators of **fluctuating asymmetry** – undirected differences between the left and right sides of different morphological structures, which were normally bilaterally symmetrical. Such differences are usually caused by errors in the course of organism development. Under normal conditions, their level is minimal, but it increases at any stressing impact, which results in increase of asymmetry.

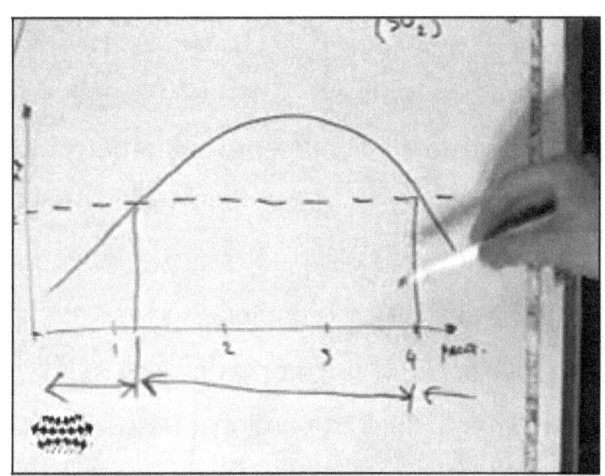

Peculiarity of development stability is that it greatly depends on general genetic changes in the organism and that is especially important when evaluating effects of radiation impact.

Assessment of fluctuating asymmetry of bilateral organisms made a good showing in determination of the total load of **anthropogenic impact**. Traditional methods evaluating chemical and physical parameters do not provide integrated idea of impact upon a biological system, whereas **bioindication indexes** reflect organism's response to the whole diversity of impact factors and they have biological meaning at the same time.

Optimal objects for bioindication of anthropogenic impact according to the mentioned procedure are plants. Animals, especially higher ones, are not right for such bioindication. First of all, they are more complicatedly organized and their development stability depends on more factors. Secondly, they are found at higher levels of pyramid of numbers and they are less exposed to pollution of soils and air environment. Finally, animals can move and they are less attached to a certain site.

Plants as producers in the ecosystem are attached to a local area all their life and are exposed to influence of soil and air environments, which reflect the whole complex of stressing impact upon the ecosystem better.

Within the **framework of the given lesson**, students are proposed to assess development stability (degree of fluctuating asymmetry) on the example of leaves taken from one of deciduous tree species found in their area. This work is not difficult in terms of techniques and required knowledge, but it is very scrupulous. Minimum of equipment is required for the fulfillment of the work: dividers, a ruler, protractor, forms for measurement data recording and counting equipment (a calculator or a computer).

General information on organization of studies

When **planning the work** it is necessary to bear in mind that its field part (collection of field material) takes insignificant share of total time required for conduction of the study. Most part of the time will be spent in the laboratory as students will take measurements and conduct estimations. Hence, it is necessary to take it into account while planning the total time of studies.

The matter is that any bioindication study is of scientific significance only in case if it is carried out not in one point and at single moment of time, but it covers several geographical points or is extended along several time periods. In other words, data of most bioindication studies is relative and "work" only upon the condition of obtaining data on its spatial or time variability (dynamics).

Fluctuating asymmetry data obtained at one time and/or in one point can be evaluated only relatively and they will indicate little if they are not compared with another point or another time period. Hence, if the study is aimed not only at teaching students according to the procedure, but also at obtaining real scientific results, it is necessary to plan to collect field material and, correspondingly, collection of data **at several points in the area** which are located at

different levels (stages) of anthropogenic impact.

If such task is possible at these studies (taking into account availability of time and "labor resources"), so it is necessary to plan

collection of material at different sites – with obviously different levels of **anthropogenic load** – as it has been done at the lesson devoted to lichenoindication (see the "Assessment of Air Pollution by Lichen Indication Method" manual, Part 5: Bioindication and Nature Monitoring).

It is ideal, if sites for conduction of these studies coincide with sites where *lichenoindication studies* were carried out during the lesson in fall. It will provide an interesting opportunity to compare two different procedures for bioindication on the example of the same sites.

Taking into account insignificant time, which is required directly for collection of field material, and large laboratory part, conduction of these studies can be carried out in bad weather (for instance, in rain).

Collection of field material

Objects under study

As it has been stated in the introduction, from a theoretical point of view studies of fluctuating asymmetry can be carried out on the basis of any bilateral (symmetrically organized) objects – both animals and plants. However, the simpler and larger the organisms, the easier is it to take measurements. Hence, leaves of deciduous tree species can serve as model objects, which are suitable for organization of similar studies. Such tree species include *poplars, birches* and *maples.*

As birch is one of the most wide-spread tree species in the central part of Eurasia, so it is proposed to use one of birch species – **common birch** (*Betula pendula* Roth.) or **white birch** (*B. alba.* L.) as the main object for studies within the framework of the work. If there are no such species in your area, the studies can be carried out on the basis of other deciduous tree species.

Time for collection

Collection of material can be conducted after intensive growth of leaves is over until the time of abscission of leaves so under conditions of the central part of Russia it corresponds with the time period from the end of May till the end of August.

Selection of plants

Collection of leaves should be conducted from trees found in roughly **the same ecological conditions** according to the level of illumination, moisture and type of a biotope. For example, one of the sites for collection should not be located at the edge of the forest while another one is in the forest.

Only even-aged plants should be used for analysis, it is necessary to avoid young trees and old ones.

Collection of leaves

Leaves are collected from 10 trees which grow close to each other, so total number of leaves collected at one site will be 100. Students should take more leaves from a site – in case if they have collected damaged leaves.

Damaged leaves can be used in the research only in the case if the sections of a leaf, where measurements are taken, are not damaged (see below). However, in order to avoid mistakes, it is recommended not to use damaged leaves at all.

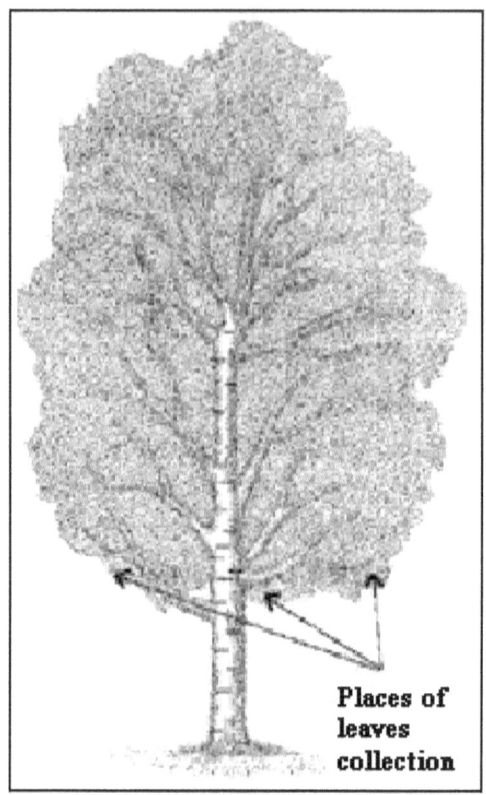

Figure 1. Places of leaves collection

Leaves are taken from the lower part of the tree crown, at the height of raised arm from a maximal number of accessible branches (Figure 1), "Places of leaves collection"). At the same time, it is recommended to try to use branches growing in different directions, conditionally – in the north, south, west and east.

Leaves are taken only from short shoots of the birch (Figure 2). Leaves should be taken of approximately the same, **average** size for the given species.

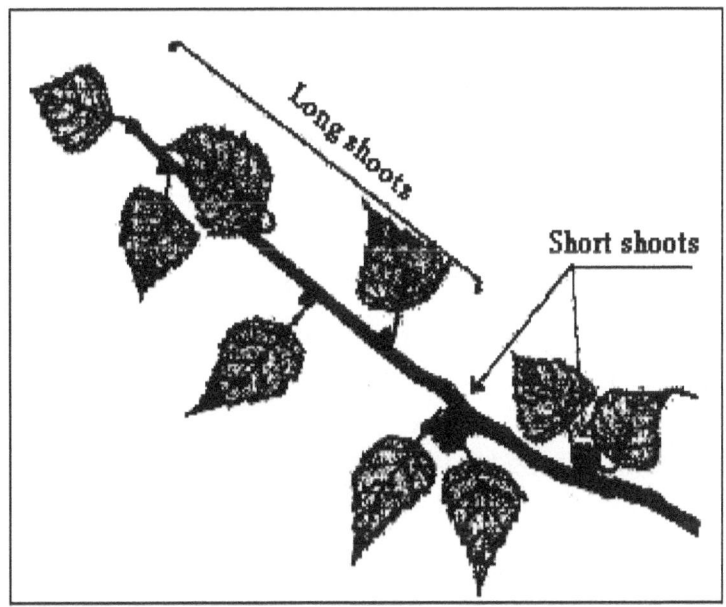

Figure 2. Different types of birch shoots.

If there are no birches in the area, where these studies are carried out, then other species of deciduous trees can be used as an object under study.

Leaves taken from one tree are tied with a string round their leafstalks and put into a plastic bag to be transported to the field studies center. Each bag (sample) is provided with a **label**, where the following information is written down: date, place of collection (it is necessary to provided detailed information on the area) and number of the site, as well as names of the author (authors) of the collection.

Laboratory processing

It is recommended to **start processing of the collected material at once** so that collected leaves are not faded yet. If the collected material cannot be processed at once, it is placed on the lower shelf in the fridge (maximal time of storage is 1 week). If leaves should be stored for a long time, it is necessary to use fixage – alcohol, diluted by 1/3 with glycerin or water.

Measurements

The following equipment is required for processing of the collected material: a ruler, dividers and a protractor. If measurements are taken by several groups, then it is necessary to check that all rulers and protractors are identical.

Let us analyze the procedure for measurements on the example of a birch leaf (Figure 3).

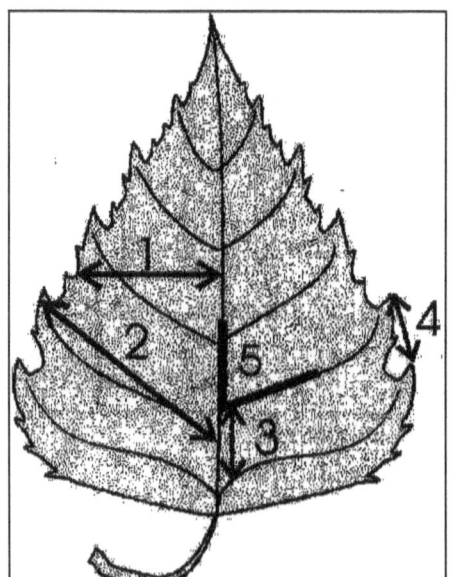

Figure 3. Parameters of the leaf

Measurement of **five parameters** are taken on the left and right sides of each leaf:

1 – **width** of a half of the leaf. In order to take the measurement, the leaf is folded across into a half by placing the top of the leaf against

the leaf base, then it is unfolded and measurements are taken along the made crease;

2 – **length** of the second vein of the second order starting from the leaf base;

3 – **distance** between bases of the first and the second veins of the second order;

4 – **distance** between ends of the same veins;

5 – an **angle** between the main vein and the second vein of the second order starting from the leaf base.

First four parameters are measured with dividers (if dividers are not available, then measurements can be taken with a ruler with clear millimeter points). The angle between veins is measured with a protractor (Figure 4). It is convenient to use transparent plastic protractors.

When measuring the angle, the protractor (position 1, Figure 4) is placed so that the center of the protractor window (position 2, Figure 4) is found at the branching point of the second vein of the second order (Position 4, Figure 4).

As veins are not straight-lined but curved, so the angle is measured in the following way: the section of the central vein (position 3, Figure 4), which is located within the protractor's window (position 2, Figure 4) is superposed with the central ray of the protractor, which corresponds with 90°, whereas the section of the vein of the second order (position 4, Figure 4) is extended to the degree values of the protractor (position 5, Figure 4) using a ruler.

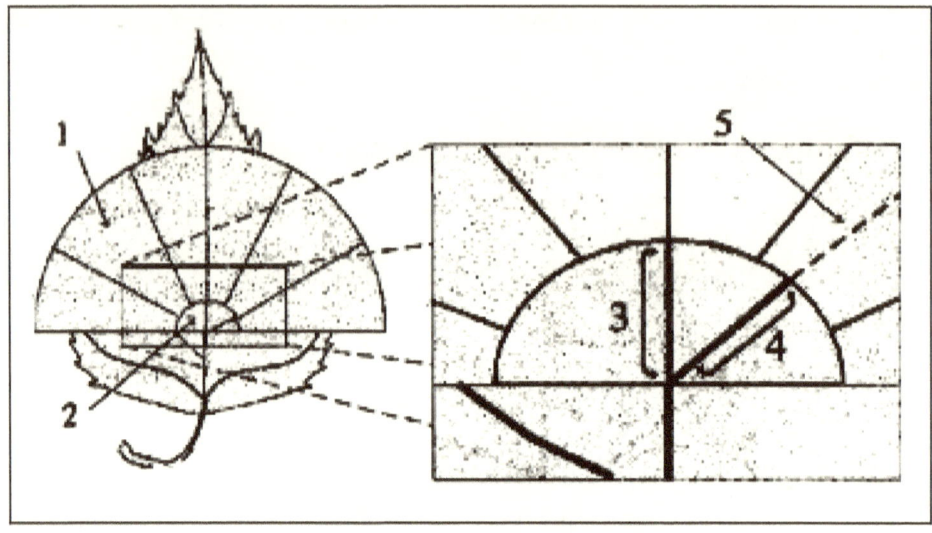

Figure 4. Measuring the angle between veins

It is advised that all leaves from one sample are measured by one

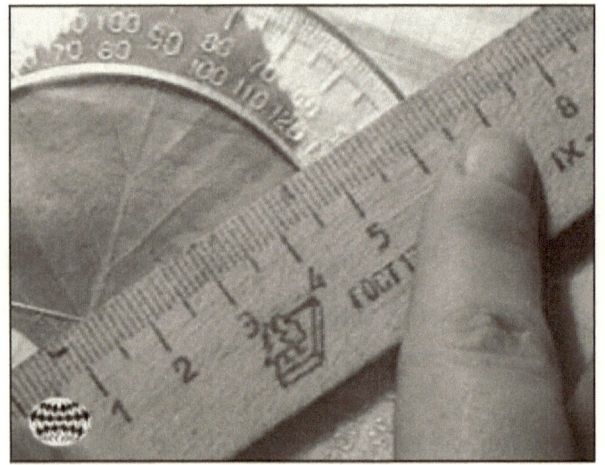

student – in order to avoid influence of subjective errors. Students should have in mind that not the absolute values of different parameters are of interest, but the difference between the left and the right halves of the leaf. That is why, special attention should be paid all the time to the technique for taking measurements of the left and the right sides of the leaf (position of the ruler and the protractor, illumination, etc.).

Measurement data is recorded in the table (see Table 1, example). It is advised to take measurements by a team of two or three students in order to speed up the process of taking the measurements. One student works with dividers (or a ruler) and measures linear parameters (parameters 1-4). The second student works with a pair of compasses and measures only angles (parameter 5). The third student records the data in the table from dictation.

The program Microsoft Excel is used for storage and computer processing of data if the computer is available.

Table 1. Values of measurements (example)

Date: _____

Place of sample collection: _____

Student: _____

№ of the leaf	1. Width of halves of the leaf, mm		2. Length of the 2nd vein, mm		3. Distance between the bases of the 1st and the 2nd veins, mm		4. Distance between ends of the 1st and the 2nd veins, mm		5. Angle between the central and the 2nd veins, degrees	
	Left	Right	Left	Right	Left	Right	Left	Right	Left	Right
1	21	20	31	29	4	5	9	9	43	45
2	20	20	32	30	5	5	8	7	40	44
3	19	21	30	30	5	4	8	9	39	41
4	17	18	27	29	3	5	9	8	46	42
5	23	20	28	31	5	4	7	7	44	41
6	21	21	30	27	4	3	8	9	39	40
7	19	19	26	29	4	4	1	9	38	42
8	20	21	29	29	5	4	8	8	41	45
9	22	18	28	31	5	3	7	9	40	42
10	24	21	21	30	3	5	9	11	42	46

Calculations

Value of **asymmetric property** is evaluated with the help of an **integral index** – the value of average relative difference divided by a characteristic (arithmetic middling of the ratio of average relative discrepancy to the sum of measurement parameters of the leaf on the left and right sides, divided by a number of parameters). Subsidiary table (Table 2) is used for carrying out calculations.

Let us denote the value of one parameter by X, then values of measurements taken on the left and the right sides of the leaf will be marked as Xl and Xr, correspondingly. Taking measurements of 5 parameters of one leaf (on the left and rides sides of it) we will obtain 10 values of X.

In the first operation **(1)** we will calculate the **relative discrepancy** between left and right values of the parameter - (Y) – for each of the parameters. In order to do it, it is necessary to calculate difference of measurement values according to one feature for one leaf, then all such values are summed up and the difference is divided by the sum. For instance, in our example, the leaf # 1 (in Table 1), according to the first parameter $Xl = 21$, whereas $Xr = 20$. Then we find out the value of Yi according to the formula:

$$Y_i = \frac{X_л - X_n}{X_л + X_n} = \frac{21-20}{21+20} = \frac{1}{41} = 0{,}024$$

The found value Yi is recorded in the supplementary Table 2, into the column 1 for the parameter.

Similar calculations are carried out for each parameter (from 1 to 5). Students will find 5 values of Y for one leaf in the result. Such calculations are carried out for each of the leaves separately and all results are recorded in Table 2.

Table 2. Supplementary table for calculations (example):

No of leaf	Parameter 1 (1) $Y = \dfrac{X_я - X_n}{X_я + X_n}$	Parameter 2 (1) $Y = \dfrac{X_я - X_n}{X_я + X_n}$	Parameter 3 (1) $Y = \dfrac{X_я - X_n}{X_я + X_n}$	Parameter 4 (1) $Y = \dfrac{X_я - X_n}{X_я + X_n}$	Parameter 5 (1) $Y = \dfrac{X_я - X_n}{X_я + X_n}$	Average relative difference per (2) $Z = \dfrac{Y_1 + Y_2 + Y_3 + Y_4 + }{N}$
1	0,024	0,033	0,111	0	0,02	0,038
2	0	0,032	0	0,067	0,048	0,029
3	0,05	0	0,11	0,059	0,025	0,049
4	0,029	0,036	0,25	0,059	0,045	0,084
5	0,07	0,051	0,11	0	0,035	0,053
6	0	0,053	0,14	0,059	0,013	0,053
7	0	0,055	0	0,053	0,05	0,032
8	0,024	0	0,11	0	0,047	0,036
9	0,1	0,05	0,25	0,125	0,024	0,11
10	0,07	0,016	0,25	0,053	0,045	0,09
						(3) $X = \dfrac{Z_1 + Z_2 + ... + Z_n}{n}$ $= 0,057$

In the second operation **(2)** students will find the value of **average relative difference** between sides of the leaf per a parameter for each leaf (Z). It is done in the following way: the sum of relative differences should be divided by a number of parameters.

For instance, for the first leaf: $Y_1 = 0,024$; $Y_2 = 0,033$; $Y_3 = 0,111$; $Y_4 = 0$; $Y_5 = 0,023$.

We will find the value of Z_1 according to the formula:

$$Z_1 = \frac{Y_1 + Y_2 + Y_3 + Y_4 + Y_5}{N} = \frac{0,024 + 0,033 + 0,111 + 0 + 0,023}{5} = 0,038$$

where N – a number of parameters. In our case $N = 5$.

Similar calculations are carried out for each of the leaves. Found values are recorded in the right column in Table 2.

In the third operation **(3)** students will calculate **average relative difference** per a parameter for the whole sample (X). It is done in the following way: all values of Z are summed up and divided by a number of the values:

$$X = \frac{\sum Z}{n} = \frac{Z_1 + Z_2 + ... + Z_n}{n} =$$
$$= \frac{0,038 + 0,029 + 0,049 + 0,084 + 0,053 + 0,053 + 0,032 + 0,036 + 0,11 + 0,09}{10} = 0,057$$

where n – a number of values of Z, i.e. a number of leaves (in our example – 10).

The obtained index will characterize the degree of asymmetry of the given organism.

There is a five-point scale of deviations from the norm, which is developed for the given index (Zakharov V.M., Krysanov E.Yu., 1996), where 1 point – is a conditional norm, and 5 points – crucial state.

Points	The value of asymmetry index
1	Less than 0,055
2	0,055-0,060
3	0,060-0,065
4	0,065-0,070
5	More than 0,07

Analysis of the results

Despite of the available scale, we will draw your attention to the fact, that it should be used carefully. It was developed for a specific area and specific objects under study, thus it is relative. You cannot say for sure that the index which has been obtained for another area and the example of other objects under study is a positive deviation from the norm.

But it is quite different if measurements were taken at different sites (areas). Then, after you have calculated the asymmetry index for each of the sites separately, we can compare obtained values and come to certain conclusions on higher or lower deviations in the certain site from the norm.

Obtained results can be presented in the form of a graph where location of test sites are plotted on the horizontal axis according to the chosen scale (according to the distance away from the pollution source) and indexes of leaves asymmetry in the given sites are plotted along the vertical axis.

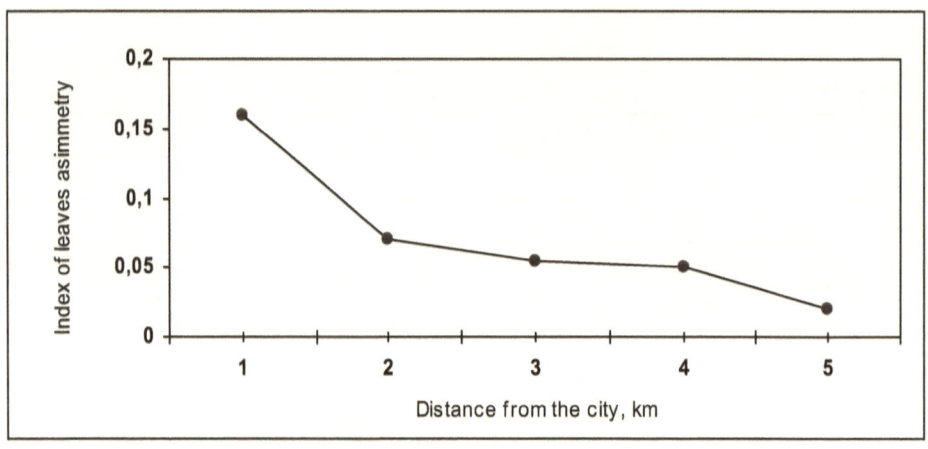

On the whole, the procedure for comparison of data obtained in different sites is similar to the procedure, which was used in the course of lichenoindication studies (see the "Assessment of Air Pollution by Lichen Indication Method" manual, Part 5: Bioindication and Nature Monitoring).

Assessment of the vital state of coniferous underbrush

This manual describes the study of the growth for coniferous tree underbrush in different types of forest plant associations. It explains the selection of biotopes for study, a procedure for measuring height and age of young trees, primary processing of measurement data and plotting graphs explaining growth success level and vital state of the underbrush.

Introduction

It is known that the rate of **reforestation** depends on a number of factors, first of all, on availability of seed and the physical properties of the environment.

Seeds of most coniferous tree species are disseminated by the wind. When seeds fall into forest litter (debris layer) under favorable conditions, they can germinate and sprout. If the specific tree species can later become the main forest layer, then such sprouts are called **underbrush** (or youth growth) from the moment of their appearance. It is widely accepted that young trees can be called underbrush until they reach 1/3 of the height of the main forest canopy.

Unlike the underbrush, trees and bushes that cannot become the main forest canopy later due to their biological features, i.e. to form a full-value forest stand are called **underwood** (understory).

A typical example of underbrush in a pine-spruce forest can be young *spruces*, *pines* and *birches*, whereas underwood is composed of *willows*, *mountain ash*, *buckthorn*, *raspberry*, etc.

The rate of underbrush growth depends on a number of factors, first of all, on the **physical conditions** of the habitat: moisture, temperature and light conditions, soil richness, diseases and insects/pests. It is practically impossible to take into account all of these factors. However, it is possible to measure the role of all the factors in the aggregate by assessing the success of growth or vital state of the underbrush under natural conditions.

This lesson is aimed at the assessment of the underbrush growth successfulness on the example of one of the most abundant coniferous tree species found in your area. Coniferous trees were chosen as objects for study as it is easy to determine the age of young trees of most coniferous species according to their appearance. It is almost impossible to conduct the same study based on deciduous trees or shrubs.

Students will need a minimum of equipment to complete the lesson: a two-meter measuring tape or a pole of 3-4 meters long, which is divided into decimeters, as well as a field diary to record the measurement data.

Selection of biotopes for study

Objectives of the study include taking measurements of the height and age of young coniferous trees under different growth conditions. Depending on the chosen tree species (let us remember that is should be an abundant species in your area), students perform an approximate assessment of its population size in different biotopes. In compliance with the standard rules for the present study (one or two days by a group of 10-11 students), it is recommended to choose 3-5 different biotopes where underbrush consisting of young trees of the chosen species is found.

Norway (common) spruce is the most widespread and suitable species for study in the conditions of the central part of Russia (in the vicinity of the field studies center "Ecosystem"). The biotopes where that species regenerates more or less successfully include a spruce forest; mixed coniferous and deciduous forest, pine forests and small-leafed forests (a *birch forest* and *birch-pine forest*). You can choose another model species in different biotopes in other areas.

Organization of fieldwork

Before the field part of the work, a group of students is divided into teams (working groups) according to the number of biotopes under study (in our case, into four teams). Each group of students is given the same task: to find a site in the biotope where underbrush grows and to measure some trees within the site.

Choice of size class of the underbrush

All trees of a young age are subject for measurement. In the course of taking measurements, students will have to measure the height of trees, so it is recommended to choose trees within the range of heights from 0.3 up to 2 meters.

The height of such trees can be measured with a measuring tape. If there are no trees of such height or there are too few of them, then students can measure trees that are 3-4 meters high, but in this case it is necessary to prepare poles of corresponding length with marked points in advance.

Selection of the site for measurements

Students should choose a **typical forested site** with an average density of underbrush within the site. They should not select dense curtains of underbrush, which are formed, for example, in so-called "windows," which are formed in a dense forest after one or several old trees have fallen. It is recommended to choose sites under the forest canopy but, nevertheless, underbrush should be found there.

An important condition for selection of the site for description is also the presence of **uneven-aged underbrush** within the site, i.e. an approximately even representation of different size classes – from the smallest trees (from 30cm tall) up to the maximum heights that can be measured (2-4 m). It will considerably enrich the informative part of the study after data processing.

Number of test trees

As in any **quantitative study**, a number of test (measured) trees should be as high as possible since the more objects are measured, the higher the reliability of the obtained data. Our experience shows that it is good to measure at least 100 trees in each of the biotopes while conducting similar studies. If this is impossible, then the minimum is 50 trees. If the sample consists of fewer trees, there is no sense in conducting the studies without an estimation of statistical validity.

Conducting geobotanical descriptions

A **standard geobotanical description** is carried out in each of the studied biotopes as the forested site, where underbrush is measured. This procedure is described in full detail in the "Study of the Vertical Structure of a Forest" manual. A form for recording the **description of vegetative cover** is attached at the end of mentioned manual. It is basically a data table with lined columns for each of the parameters of the description of the environment. The form is developed for the simplification of record keeping and the standardization of described physical parameters of the environment. The forms are filled in directly in the field on the study site.

The primary attention during the description should be paid to **crown density** (let us remember that this is the ratio of the space covered by tree crowns and open sky) and height of the forest stand. A procedure for determining crown density and tree height is also given in the three above-mentioned manuals.

If an instrument that allows you to measure **illumination** (a luxmeter) is available, then it is interesting to measure this parameter at sites under description: it will make the research better and help to formulate conclusions more clearly.

Underbrush measurements

The **main part of the fieldwork** consists in taking measurements of young trees. All trees that can be measured within the chosen sites in the forest should be measured. Two measurements of each separate tree are taken: its height, with the help of a measuring tape or a measuring stick (to within 10cm) and its age.

Two or three students measure height: one or two students measure the tree with a measuring tape, another one records the measurement data. When measuring the height of trees with the help of a measuring pole, one student places the pole next to the tree and another student goes to a certain distance away from a tree and looks at the pole: specifically, which mark on the pole corresponds with the top of the tree.

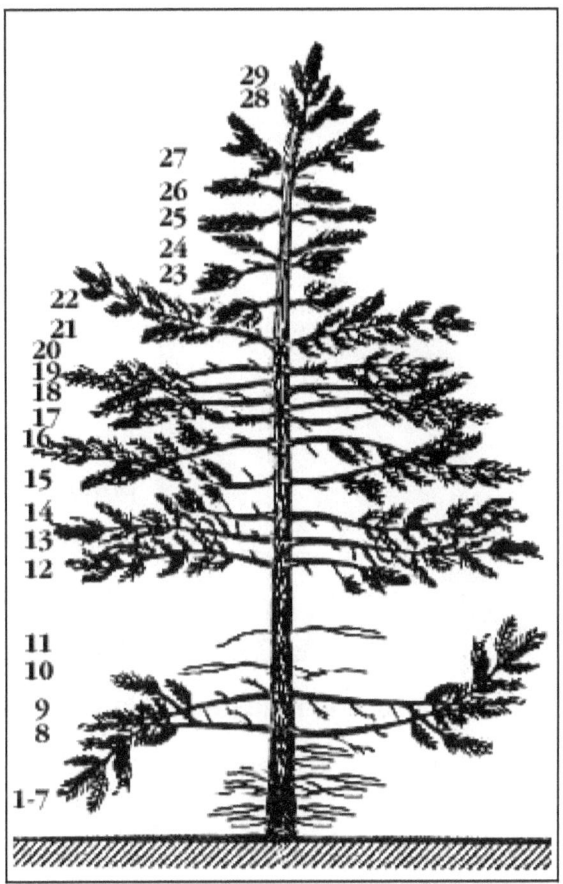

At this time, one more student estimates the **age** of the tree. The age of young trees is measured according to branch clusters called whorls. The difficulty is that most tree-like coniferous plants have branch clusters on their trunk when they are young – these are places where branches grow in clusters. When looking at the tree, the branch clusters are well seen.

Branch clusters of different tree species are more or less distinctive and they stay on the tree trunk sometimes until a tree is 60 or more years old, in particular, under unfavorable growth conditions, for instance, in thick groups of trees (plantations). It is especially characteristic of pine trees, fir trees and larches. Living branches of spruce located in the lower part of the crown fall down early and are replaced by dry twigs, particularly in the lower part of the tree crown near the ground.

As a tree grows very slowly in earlier years of its life and branch clusters belonging to that period are preserved badly, so it is necessary to add 5-8 years to the number of years estimated on the basis of easily noticeable branch clusters – this is how the absolute age of the tree can be roughly estimated.

When estimating the age, students should have in mind that distances between branch clusters, i.e. annual growth of the tree, can change greatly year to year. One should not expect that all the distances among branch clusters should be equal. One more peculiarity is that the annual growth becomes longer while approaching the top of the tree, as it starts to grow faster as it gets older.

Measurement data is recorded in the field diary or in a notebook in the form of a table, where a serial number of the tree corresponds with two values – height and age.

Data processing

After all field materials are processed, this study should result in **plotting graphs** that show the dependence of tree height on its age in different biotopes. Preliminary data necessary for plotting graphs is prepared on the basis of field records.

At the first stage, all field measurements are grouped according to **height classes,** and the average age is estimated for each of the height classes.

The number of height classes is chosen arbitrarily; it is advised that there are about ten of them. For instance, if the shortest tree in the measurement site was 20cm high and the tallest one was 3m, then the number of height classes chosen was ten, as follows: less than 30cm, from 30 to 60cm, from 60 to 90cm, 90-120, 120-150, 150-180, 180-210, 210-240, 240-270, and 270-300cm.

Students then calculate an **average age of tress** belonging to the given height class. In order to do this, they make an intermediate table, where they write down their data on the age of trees in a certain size class.

Size classes, cm	Age, years	Number of trees	Average age, years
Less than 30	6,6,7,3,8,9,6,7	8	6,5
30-60	10,6,8,9	4	8,25
60-90	13,13,14,15,12,10,18,17,14	9	14
90-120	21,15,20	3	18,7
...	...	...	...
270-300	38,32,34,32	4	34
Total		100	

When students have written down the height of all measured trees, they estimate the **average age of trees** belonging to that height class (the sum of all ages is divided by the number of trees) and write it down in the right column of the table.

Calculation data is plotted on the graph (Fig.1):

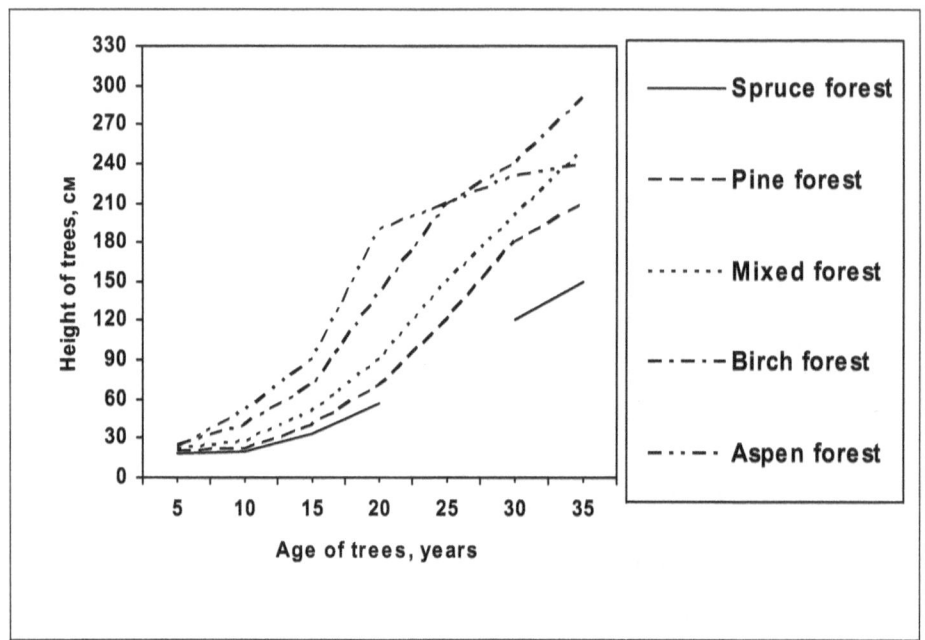

Fig. 1. Dependence of the height of the coniferous underbrush upon their age for different biotopes.

The height of trees is plotted on the vertical axis (axis of ordinates) and the age of trees is plotted on the horizontal axis (abscissa axis). If ten height classes are chosen, then there will be ten points on the graph, which is joined with a line. The steepness of the line indicates the amount of successful tree growth, i.e. it reflects vital state of the underbrush.

The procedure for **estimation of average age** and **graph plotting** is conducted separately for each of the studied biotopes. Moreover, a uniform scale of age classes (in our case, every 30cm) is used in intermediate calculations of the average age of trees for all the biotopes. If three of the certain age class are not found in the specific biotope then a point which corresponds with the given age class will be lacking on the graph.

Data on dependence of the height upon age for different biotopes is plotted on the same graph. In an example with four biotopes there should be four lines in the graph. Perhaps they can be of different lengths (for instance, there are no very high or very short trees within the given biotope) or interrupted (there are no trees of one of the size classes). Students should not join points that are lacking on the graph with one line (extrapolate)(Fig. 1).

Lines characterizing different biotopes are marked in different conventional symbols (for instance, in different colors or line styles).

Curves, obtained for different biotopes, are analyzed, i.e. the vital state of underbrush in different biotopes is compared. The higher the line is on the graph, the higher the vital state of the underbrush in the given biotope (the larger are trees of the same age). The steeper the curve, the more active the growth of the underbrush, i.e. the general conditions of underbrush growth are more favorable.

The same data can be presented in a form of color diagrams (Fig. 2):

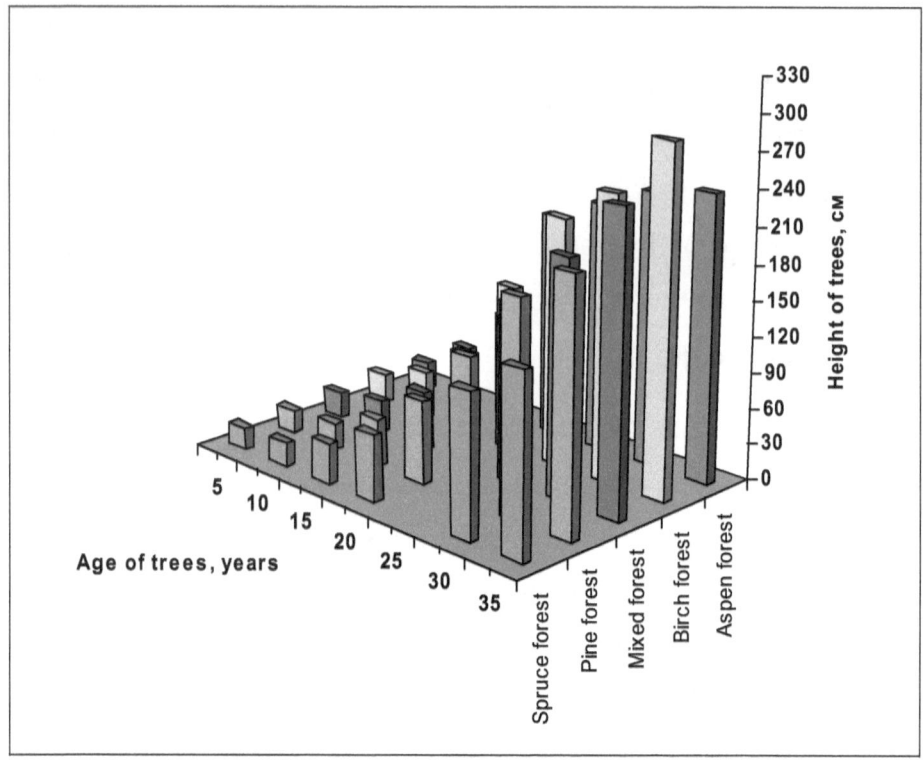

Fig. 2. Dependence of the height of the coniferous underbrush upon their age in different biotopes.

On the basis of obtained data students should try to answer **the following questions**:

1) Does the vital state of the underbrush differ in different studied biotopes?

2) What are the differences in the total size of trees or in the rate of their growth?

3) What are possible causes of differences in vital state of the underbrush in studied biotopes (moisture, light, soil richness, intra- and intraspecific competition, influence of people, etc.).

Forest Vegetation Description Form № _____

Date: _____ Authors: _____

Administrative and local position: _____

_____

Position in the relief: _____

Surrounding plant associations: _____

Described area (м х м): _____

The name of plant association: _____

Arboreal and Bush Layers	Crown Density	Formula of the Forest Stand	D (1,3)	H (fs)	H (ca)	Age
Old and Adult Forest Stand						
Young Growth					-	
Understory (Bush Layer)					-	

D (1,3) - diameter of tree trunks measured at chest height (~1.3m) in centimeters, H (fs) – average height of the plants in meters; H (ca) - the height of crown attachment (an average height at which lower living branches of trees are found) (only for the forest stand).

Herbaceous-Shrub Layer

Hillocks: Depressions:

Moss-Lichen Layer

Hillocks: Depressions:

Complex Environmental Assessment of Human Impact on an Area

This manual contains a procedure for human impact study, which

allows us to quantitatively evaluate the degree of disturbance and the influence of human activities on the main landscape components: relief, vegetation and soils. This study is based on the route counting of local objects of human impact. The procedure is applicable for forested zones.

Introduction

When conducting complex environmental expeditions, practical work and environmental camps, young environmentalists and their leaders often face a task of the **quantitative** estimate of total human impact on the surrounding environment. This manual provides a simple procedure for **conducting a complex route count of objects of human impact on the area.**

This lesson and the study procedure itself are developed primarily for research in the forested zone – in the vicinities of a city, in a Forest Park or in a large city park. This procedure is not applicable for small-forested areas less than ½ square kilometer. In this case it should be replaced by mapping the objects of human impact.

This study can be carried out by a group of students who do not have skills or experience in complicated field research work. This work does not require special knowledge of the names of animals or plants species (except some main tree and shrub species). Only the well-coordinated work of the whole group is needed en route.

Methods (Procedure) of route counting

General information (Basic principles)

Route counting of objects of human impact (HI) is aimed at discovering the local forms of human impact on the environment, which can be revealed and described based on cartographic documents. This applies, firstly, to anthropogeneous damages of microrelief, soil, vegetation and wildlife. While counting of objects of human impact, the main types of plant associations are also registered as a foundation of the studied landscape.

Route counting allows us to obtain **quantitative data** on the anthropogeneous load on the local environment, which can be compared with analogue data collected in other areas and in other regions.

In order to compose valuable characteristics of HI on the selected area it is advised to locate a **counting route**, which crosses **typical** habitats within the boundaries of the selected area.

Counting is based on the **method of the "random"** route, which is a straight-lined route (without using roads) chosen with the help of a compass. The route can be straight-lined and closed-looped (rectangular or triangle, but not circular) returning to the starting point.

One should bear in mind when calculating the total length of the route, that longer the route is, more precise (valid) the data is. The total length of the route should not be less than two to four kilometers.

List of objects for route count of human impact objects and environmental conditions

In order to make counting easier, all objects of human impact are divided into three groups: 1) "measurable" objects: they have different sizes and their length can be measured. 2) "Un-measurable" objects: these objects usually have a "standard size," and 3) "spot" or "point" objects: they are equally distributed in the area as a rule.

"Measurable" objects are the following:

a) Anthropogeneous forms of microrelief: irrigation ditches, pits, rain rills and gullies that are artificial in origin, embankments, bars, bumps, terraces on slopes etc.;

b) Soil and herbaceous vegetation damages caused by humans - vast areas damaged by heavy vehicles, highways and country roads, paths, pits, grooves, garbage dumps, cattle tracks, large soil trampling areas

c) Artificial water bodies and waterways - ponds, swamped areas, soil-reclamation and other canals, drainage systems and so on;

d) forest anthropogenic damage - forest openings (various causes: forest management cuttings, cuttings for electric power and communication lines, forest roads and pipelines), felling grounds and burned-out forest areas.

Besides the description of human impact, the **main types of plant communities (associations)** along the route should be also described. In the Moscow region, for example, the following communities can be seen: spruce, pine, small-leaved, mixed, broad-leaved, alder forest, inundated brushwood, meadows and so on.

The group of "un-measurable" objects consists of electric power lines, communication lines, pipelines and other objects, that have standard width.

"Spot" or "point" objects are the following: **Household garbage** lying on the ground - paper, plastic, bottles, cans, etc., anthropogeneous damages of soil and herbaceous vegetation - **campfire sites** (less that 1 m in diameter and more than 1 m in diameter); anthropogeneous damage to trees - dry trees and top-drying trees, cut trees (stumps), wind-fallen trees, trees with mechanical damage to the trunk (trunk wounds and traces of charring); encounters with **synanthropic (associated with people) animals** (hooded crow, rook, jackdaw, starling, sea-gull, stray cats and dogs).

Counting procedure

Counting is carried out by a group of students (5-12 people) that walks along the route strictly according to the compass. Strict attention should be paid to the walking along the route exactly; any route deviation discredits the entire procedure. The initial direction is voluntarily selected at the starting point, usually in the direction of the main forest massifs in the area.

When counting, all primary data is recorded in field diaries of the counting group members. Examples of data recording are presented in Fig. 1.

Many tasks are performed during counting, and several counting recorders are responsible for each activity:

One person ("navigator") leads the group according to the compass, and if possible, maps the route (this is not obligatory). His or her task is to ensure strictly straight-lined movement along the previously chosen direction (according to the azimuth). He or she is also responsible for changes in direction and bringing the group back to the camp.

The second member of the group ("distance recorder") counts the distance covered (in steps or with the help of pedometer). For further calculations this person should estimate precisely length of his or her step (or a pair of steps). In order to do this, the distance recorder

should count the number of his or her steps to reach the distance of 100 meters, which is measured with a tape measure. When estimating a distance in steps without using a pedometer, it is easier to count not each step, but a pair of steps, and record each hundred pairs of steps in the field diary.

A group of students consisting of three people, for example, measures all "measurable" objects AB with a tape measure, which are met en route. Note that their length is measured **strictly along the straight line** of the counting route (not perpendicularly). Data on each object is registered in the field diary with a semicolon (see Fig.1). If there is an "insurmountable" object (a quarry, a ditch or a water reservoir) along the straight line, its length is measured by eye. The distance required for making a detour around the object is not included in the distance estimations. Distance measuring restarts from a point where the straight-lined trajectory of the route comes out on the opposite side of the objects.

The second group of students (for example, 2 people) records all plant associations along the route, making a brief description of them on the way. In order to measure the length of each association, students from this group ask the "distance recorder" the number of steps at the points where one plant association is replaced with another. Boundaries of the plant associations are recorded in the number of steps from the starting point. Records are written in the following way (for example): *spruce forest - 0-368; 534-887; pine-spruce forest - 368-534; mixed forest - 1230-1456* and so on.

The third group of counting recorders (5-6 people) counts all "spot" or "point" objects within the standard, chosen width of the counting band (strip).

The width of the counting band can vary from 10 to 40 meters (5 - 20 meters to each side from the straight-lined route) depending on the view of the vicinity and the counters' diligence. The denser the forest is, the more narrow should the counting band should be. The wider

the band, the more difficult it is to count objects; however, a larger area can be covered by the same distance (and the more accurate the collected data). Once the width of the counting band is chosen, it should be strictly followed - none of the objects outside the boundaries of the counting band can be included in the records. Before counting, participants should be trained in how to measure the counting band by eye using multiple checks of its width with a tape measure or in steps.

It is advisable to allocate responsibilities within the group of "point" objects recorders depending on the number of objects encountered. For example, two students can count trees with dry tops and those with damage to the trunk, two can count household garbage and campfire sites and one can count stumps and wind-fallen trees as well as synanthropic animals.

Data is recorded using the **"accumulation" method** according to the "symbol" system: . - 1, .. - 2, ⁚ - 3, ⁝⁝ -4, ⊔⊔ -5, ⊔⁚ - 6, ⊔⊔ - 7, ⊓ - 8, ⊠ - 9, ⊠ -10. This system allows students to add new numbers to existing ones easily and to calculate a resulting sum quickly when transferring data from a field diary to a final census sample (finished squares or "envelopes" ⊠ correspond to 10)(Fig.1).

Lastly, one person counts all "un-measurable" objects that **cross the route**. In this case, objects are also recorded according to the library method (Fig.1). Objects that the route does not cross (even if they are located close to the route) **are not counted**.

Fig. 1. Data 23.09.98. Route #1 (see the map)
Route length: 2 260 m. Counting time: 3 h 20 min

Measurable Objects

No.	Measurable Objects	Length
1	spruce forest	680; 250; 780........
2	forest openings	8; 12; 10; 2; 34; 29........
3	mixed forest	230......
4	forest clear cuts	34; 23.........
5	forest roads	2; 5; 12..........
6	paths	1; 4; 6...........
7	cattle crossing	56............
8	etc.	

Non-Measurable

No.	Non-Measurable Objects	Number
1	power lines	..
2	connection lines	⊠..
3	gas pipes	⊠
4	domestic animals	⊠⊠..

Spot Objects Within Counting Zone (20 meters)

No.	Non-Measurable Objects	Number
1	campfire site < 1M	⊠⊠..
2	campfire site > 1M	⊠⊠⊡∴:
3	damaged trees	⊠⊠⊡∴:
4	dead trees	⊡∴.
5	dying trees	⊠⊡∴.
6	fallen trees	⊠⊠⊠⊠⊡∴:

Fig. 1. Example of record keeping in the field diary for route counting of human impact objects and environmental conditions.

Processing the data

When the students come back to the field center, each group processes all collected primary data from the field diaries according to their objects. All data are totaled and estimated separately for each object and then they are entered into the corresponding tables.

Count data are calculated in the following way:

For **"measurable" objects:** when the counting is over, the total length of each object along the route is estimated (in meters or in steps and then the equivalent in meters). When the total length of the route is known, the percentage or share of each object of HI is estimated for the studied area (in % from total route length).

For **"un-measurable" objects**: after the counting is completed, the total number of times each object is encountered en route is calculated. This value then is estimated based on the total length unit (for example, per one linear kilometer of the route).

For **"spot" objects**: the total number of objects within the band is calculated. The total studied area is calculated by multiplying the total route length by the width of the counting band. Then the "density" of each object of HI per square unit can be estimated (per ha or sq. km.).

Presentation of results

Results of the study are presented in the form of a table with the title "Objects of human impact in the area (-s) (Name of the area (-s) where the counting was carried out)". All found objects of human impact are listed in the left column of the table, and their representative values for the studied area are presented in the right column.

For measurable objects, their representation is calculated in % from total length of the route; for "un-measurables" – in units per 1 km of the route; for "spot" – density in units per 1 sq. km.

The title of the table should also include the total length of the route, total counting time and width of the counting band.

A scheme of the route drawn from the map of the area should be submitted with the table of results. Boundaries of plant associations and very large objects of human impact (burned-out forest areas, clearcuts, quarries) can also be marked on the map.

If more than one group of trained students are available, or if you have several habitats with contrasting human impact (for instance, in close and distant city vicinities), it is possible to conduct counts in each of the habitats separately and then compare the results (quantitative composition of objects of human impact and their representation). Conclusions on the difference among the areas according to the intensity of anthropogeneous load can then be made.

The study of water invertebrates in a local river and assessment of its environmental state

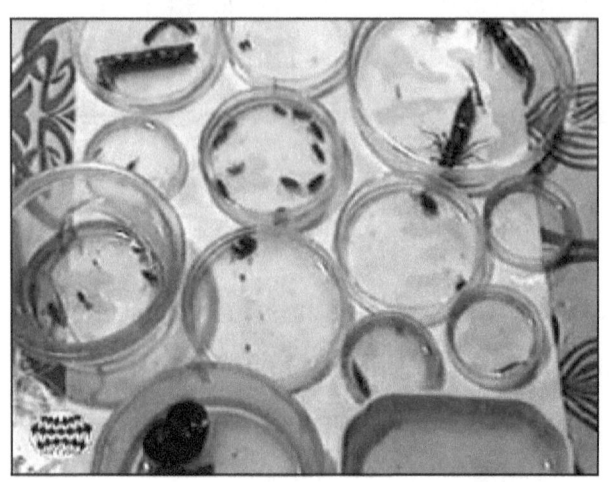

This manual is devoted to the study of macrozoobenthos - invertebrates that inhabit the bottom of bodies of water. The main idea of this lesson is to study these invertebrates in a local small river or stream and to determine water quality based on species composition and representation of various collected organisms.

Introduction

Zoobenthos (from bentos – depth) all the invertebrates that inhabit the bottom of bodies of water (or the **benthic** zone), aquatic vegetation (or **phytal**) as well as other substrates including those associated with **hydroengineering** facilities.

The largest benthic specimens, with a body size of more than 2 mm, are called **macrozoobenthos**. The following make up the category Macrozoobenthos: worms (**planaria, oligochaetae**, *leeches, round worms*), **mollusks** (**gastropoda and bivalves**), **Crustacea** class (**amphipods, isopods, decapods**, etc), **arachnids**, and insects (*midges,* **geleides**, *May-flies, stoneflies, caddis-flies, dragon flies* and so on), etc.

Many of these living organisms also dwell in the water column (**pelagic** zone); they include insects, **Crustacea** class (**mysidacea, pallacea** and others), *spiders*, etc.

The lives of many other bottom organisms can also be connected with the water surface, i.e. surface film (**neistalic** zone).

According to its functions, macrozoobenthos represents an important part of the **heterotrophic component of water systems**. Macrozoobenthic organisms are particularly involved in the processes of converting outside energy sources (plant and detritus material) into available energy within the system.

Changes in the species structure of *biocoenosis* correlating with level of water pollution have attracted **hydrobiologists'** attention for a long time. The high **stenobionicty** (demands for certain environmental conditions) of a number of species, formation of complex multi-component systems, attachment to certain types of substrates, relative low motility (mobility) (in comparison with fast-spreading pollutants) allow the use of zoobenthic organisms as indicators for determining human (**anthropogeneous**) impact on aqueous ecosystems.

Different methods of water quality assessment, numerous articles and publications as well as the data of long-term observations within the hydrological network of environmental monitoring prove the importance of benthos for determining characteristics of water quality. Because macrozoobenthos specimens are relatively large, their detection and determination by young ecologists (environmentalists) is made easier.

Procedure for studying macrozoobenthos

General information

The study of macrozoobenthos and fulfillment of this educational task can best be carried out at a **nearby small river or stream** with a slow current.

An ideal location for studying zoobenthos is a river valley, which is not too wide (5-20 meters wide) and not too deep (up to 1.5 meters deep), or a relatively flat section of a mountain river with a slow flow and well-developed high aqueous vegetation. A fast mountain river

with a coarse-silted bottom and absence of high aqueous vegetation, or reservoirs with stagnant water are less suitable for this research.

In order to make sample collection easier it is recommended to choose a shallow river section to collect samples at different places by walking in the river (not from a boat or a bridge). Periods of sample collections are limited by the **seasonal life cycles** of benthic organisms. These life cycles occur year round and are not limited to the period from July to November.

Sample collection

Selection of sites for sampling in the river is the starting point of all hydrobiological studies.

For the purposes of this task, an **average river section** should be chosen. The best area will have favorable oxygen conditions, such as shade and high water vegetation. It is not advisable to collect samples at where ground water discharges, at stagnant sections, or other unfavorable river sites. These areas will not give a true indication of the entire river system.

Samples collected from the *abyssal* part of a river are also not suitable, because they may not characterize water quality but rather, the pollution of bottom sedimentation, which can greatly differ in chemical composition from the water in the river as whole.

For these studies, mountain and piedmont rivers with rocky-pebbled bottoms are best. If these conditions are unattainable, (or for river valleys) samples should be collected from submerged *macrophytes* (high water plants). If there are no enumerated substrates, or in the case of a water level increase, samples should be collected from the merged *terricole* or submerged vegetation. If such vegetation is also not present, then samples are collected from soft soils – sand, clay and silt.

To conclude this section, we can once again outline that samples according to the objectives of this lesson should be collected **in average** (on all parameters) river sections and, certainly, **in different parts** of the river.

Sampling technique

The most convenient and universal tool of macrozoobenthos collection is a scraper, which is a metal frame with a cutting edge with attached gauze or netted bag placed on a stick (also known as a D-net.)

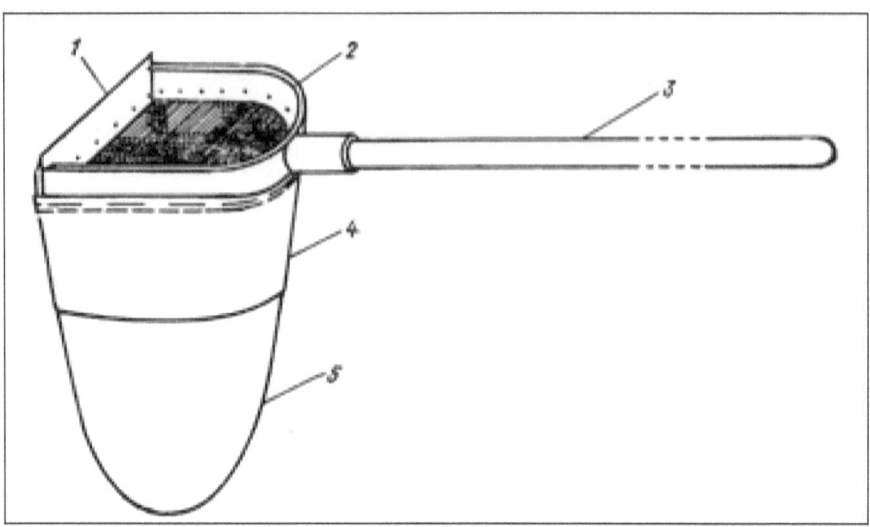

Figure 1: Scraper: 1 – cutting edge; 2 – frame; 3 – handle; 4 – cotton fabric part of the washing sieve; 5 – part of the sieve made of kapron.

The scraper allows the student investigator to collect both **qualitative and quantitative samples from all types of substrates**, including such specific substrates as the submerged overgrown sides of *ferries, walls of hydroengineering facilities, bridge piles* and so on.

The sampling technique using a scraper has some peculiarities. When collecting samples in the river, the scraper should first be placed down the **watercourse** (downstream) from the substrate from which the samples are to be collected. In that case, organisms together with suspended soil particles or substrate fragments will fall into the scraper's screen with the water current. In fast rivers, stir the bottom with a foot; a person should move with one side forward and place the scraper downstream.

On rocky substrates the organisms should first be washed into the scrape with a gentle movement of a hand from the rock surface, then the scraper should be turned upside down and it should smooth the

bottom surface. When a large cluster of algae or macrophytes fall into the scraper, without taking them out of the screen they should be shaken in the water and then removed. Large pebbles that fall into the screen should be removed after the screen is carefully examined and all organisms are taken out with the help of forceps.

When samples are collected from plant thickets and filamentous algae, they should be shaken in the screen of the scraper, submerged into the water and then they should be once again examined in order to collect attached organisms. When collecting samples from dense thickets of microphytes, the scraper should be pushed into the thicket; then the thicket should be "mowed" with sharp, energetic movements.

After each collection of samples, the scraper should be taken out of the water and its contents should be put carefully into a container or a dishpan filled with clean water from the river. The screen should be turned inside out. All organisms seen with the naked eye should be collected (with fingers, with the help of pincers (forceps), a spoon, etc. – depending on the size) and placed into a glass container with a wide neck for sample storage and transportation to the field center. Due to active movements even small organisms are well seen in a white pan. NOTE: Specimens can be separated in the field (old ice cube trays work well), identified and counted if they are to be returned directly to the water.

When collecting samples from sandy, soft clayey bottoms or silt the scraper is immersed into the bottom several centimeters deep and the bottom surface layer is cut out with a cutting edge with a scraping movement. The scraper's movements are directed up the watercourse. The method of **elutriation** is applied for sampling from

such substrates. First, soil collected in the scraper is washed out directly in the scraper, which is immersed several times and taken out of the water, and then is transported to a bucket or a bowl with water. Then the soil is stirred several times with a rotary movement. Suspended particles together with organisms are poured into a previously rinsed scraper after each stirring and then into a pan or a container with clean water. Taking into account the low inhabitance of sandy bottoms, this operation should be repeated several times. In order to avoid rubbing the organisms with coarse sand particles and other trauma, the stirring should be done carefully with smooth, gentle movements.

For the purposes of this lesson, all organisms collected at different river sites can be placed into the same glass jar.

Accompanying description of the river and samples

Before starting to collect benthos samples, the **riverside zone should be examined**. All soils should be observed at about 50 meters up the watercourse and 50 meters down. The typical appearance of the riverside zone should be recorded directly at the sampling site and a map of the area sketched. In addition the student researchers should record: 1) number of the sample, 2) date and time of sampling, 3) name of the river, 4) location of sampling site (map).

The following **information** should also be provided: 1) water and air temperature at the time of sample collection, 2) weather conditions on the day of sampling and during previous days (retrospective information on weather conditions helps to explain the occurrence of possible abnormal findings). It is advisable to record a visual description of hydrological parameters of the river in the field register: 1) stream velocity, 2) color, 3) water transparency, 5) degree of the riverbed filling.

In order to make recording easier data can be recorded on the **Exploratory Description Form** (Table 3 at the end of this manual).

Information concerning the benthos itself should include: 1) substrate, where the samples were collected from, 2) distance from the bank, 3) depth of sampling site, 4) number of scraping movements (one scraping movement is a conventional unit of sampling area expressed in distance covered by the scraper in the bottom.) It is convenient, for example, to move a distance of 50 cm in a soft bottom (which is covered with a cutting edge) for one "scraping movement." In this case, the width of the cutting edge should be recorded as well. The description of the river is concluded with "notes, "where observations of life in the biocoenosis (such as first flight of insects, abundance of empty shells of mollusks, unformed state of the biocoenosis, etc.) are recorded.

Sample processing and species determination

Analysis and species determination should be carried out during the same day, while the organisms are still alive. After returning to the lab, the samples from the glass containers are poured out into white pans filled with, preferably, water from the same river (two to

three liters of water should be brought from the river in a bottle).

Then all collected organisms are sorted into Petri dishes. Organisms of the same species (at least according to their appearance) are placed in one dish.

The collected animals are **identified** using field guides or other visual/descriptive keys. A binocular microscope may be helpful. It is advisable that each student gets training in species determination and draws at least one organism.

When determination is finished, a total list of collected organisms is written. It is not necessary to determine correct species name; it is important to register the presence or absence of the main **indicative** groups of organisms, which will be used later for the assessment of environmental status of the river.

Assessment of the environmental state of the river based on biological index

The biological index method of estimation developed by F. **Woodiwiss** in 1964 is the most widely used method within the system of environmental monitoring. It applies the evaluation of water quality according to zoobenthos parameters.

The method is based on a relationship of simplification of **biocoenosis** taxonomic structure in correlation with an increase of **water pollution level** (owing to the absence of indicative taxons at achieving their tolerance limits) with a simultaneous decrease of diversity of organisms that belong to the "Woodiwiss's groups" (Table 1):

Each species of flatworm	Fly larvae (except midges and buffalo gnats)
Oligochaetae class (except Nais genus)	Midges (except *Chironomus thummi*)
Nais genus	Beetles
Each species of leeche	Alder and snake flies
Mollusks	Each family of caddis-flies
Crustacea class	Buffalo gnats
Stoneflies	Bugs
Mayflies	Larva of *Chironomus thummi*

When calculating the total number of Woodiwiss's groups, the presence of at least one specimen from the specified groups in the sample accounts for one point.

Among these 16 groups of organisms, Woodiwiss determined six indicative taxons. The presence of these taxons in the studied reservoir, together with presence of other animals (biodiversity of the benthic community), indicates degree of reservoir purity (Table 2). These groups were identified on the basis of a large database collected by the author.

According to this method, there is no necessity to determine exact species – the determination should be limited by the taxon specified in the table. For some of the taxons (*mayflies, stoneflies, caddis-flies*), only the fact of their species presence or absence is taken into account. Attitude to different species is determined visually according to organism appearance.

The presence of at least one specimen of the corresponding taxon is considered as its presence in the water body.

The working scale for calculation of biological index based on the presence of Woodiwiss's groups is given in Table 2:

Indicative taxons	Species diversity	Number of Woodiwiss's groups in the sample				
		0-1	2-5	6-10	11-15	16+
Stonefly larvae	More than one species	-	7	8	9	10
	Only one species	-	6	7	8	9
Mayfly larvae*	More than one species	-	6	7	8	9
	Only one species	-	5	6	7	8
Caddis-fly larvae **	More than one species	-	5	6	7	8
	Only one species	-	4	5	6	7
Gammarus	All above listed taxons are absent	3	4	5	6	7
Hog slater	Same	2	3	4	5	6
Tubiphicides and Chiromonuseae larvae	Same	1	2	3	4	-
All above listed taxons are absent	Some organisms which require lower oxygen levels can be present	0	1	2	-	-

*- Except *Baetis rhodani*
** - Including *Baetis rhodani*

When working with the table:

One should move from the top to the bottom of the left column of the table, determining if the indicative organism marked in the column is in the sample. The first organism found in your sample will be indicative – it will determine the degree of water purity. Then it is not necessary to move further down the table.

If some stoneflies, mayflies or caddis-flies are present in your sample then you must determine if you have only one species or several (according to their appearance).

If none of the stoneflies, Mayflies or caddis-flies are present in the sample then you have to move further down the table until you meet some indicative organisms which are present in your sample.

Calculate the number of Woodiwiss's groups in the sample (according to Table 1). Find the value of the biological index where the line of found diversity crosses the column of the number of Woodiwiss's groups, which corresponds to your sample.

This will be your indicator of relative water quality (purity) in the river – biological index. The higher the index is, the cleaner the water is. Biological index is a relative indicator and is measured from 0 (very dirty water) up to 10 (pure water).

Example:

Assume that two species of flatworms were found in the sample (they present two Woodiwiss's groups according to Table 1), as well as several species of Oligochaetae (all Oligochaetae species belong to one group), three species of leeches (3 groups), mollusks (1 group), Crustacea class (1), several species of mayflies (1), bugs (1), fly larvae (1), one family of caddis-flies (1) and Alder and snake flies (1).

We estimate number of Woodiwiss's groups (using Table 1); in our case it is equal to 13.

Then we find the "highest" taxon according to Table 2. It is mayflies (there are no stoneflies in the sample). We have several species, so we take the upper line. At the crossing of this line with the column "11-15," we find the biological index of our water body to be 8.

Presentation of results

A **consolidated table should be compiled** based on study results. The title of this table is "Macrozoobenthic organisms of the river and results of assessment of water purity according to indicative taxons."

The list of found organisms (taxons at the level of Woodiwiss's groups) lays the basis for the consolidated table together with drawings of each taxon (various species can also be shown).

At the top of the table, indicative **taxons** (from Table 2) should be recorded starting from the highest and then other organisms that belong to Woodiwiss's groups (from Table 1) and at the bottom of the table, a list of found species that do not belong to Woodiwiss's groups.

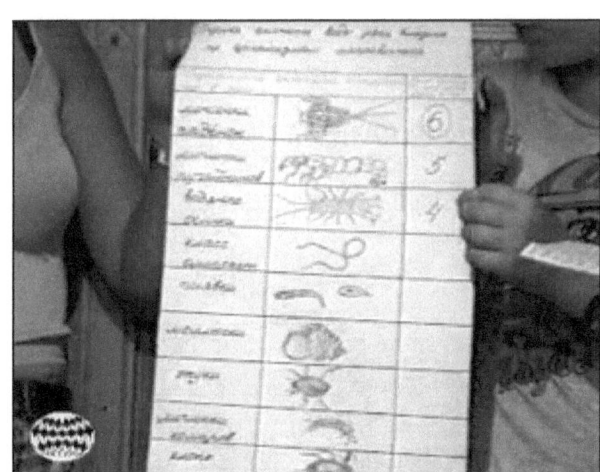

To the right against the names and drawings of the indicative organisms, corresponding biological indexes should be filled in (from Table 2). The highest (and maximum according to the index) will characterize the degree of relative purity of water in the studied river.

Additional information on river description and data on sample collection (it is recommended to mark sampling sites on a map of the river) should be supplied as an Appendix to the main table.

Woodiwiss, F.S., The Biological system of stream classification used by the Trent River Board, Chem. and Ind., 1964, vol. 11, pp. 433-447.

Table 3.

Exploratory Description Form for a Water Body № _____

1. Date of observation _____ (day, month, year)
2. Weather conditions _____
   (air temperature, cloudiness, wind strength, precipitation, snow cover and ice)
3. Type and name of the water body_____
4. Location of the studied area (observation site)_____
_____
   (administrative district where the observation site is located and its distance from the nearest human settlement)
5. Vicinity description (description of the surrounding environment) _____
_____
   (village, town, city, forest, meadows, agricultural lands, etc. and their short description)
6. Morphometrical features of the site _____
_____
   (width, average depth, stream velocity, shore type, slope of the bottom)
7. Riverside water vegetation (dominant species)_____
_____
8. High aquatic vegetation (dominant species) _____
_____
9. Description of the river bottom soil and shore soils _____
_____
   (stone type, stone-sand type, sandy soil, silt-sandy type, sand-silty type, silty type, clay type)
10. General description of water:
a) water temperature: near a bank _____ , far from a bank _____ ,
at a depth of 1 m _____ b) water color _____
   (blue, green, yellow-green, green-yellow, yellow, brown-yellow, or brown)
c) water transparency _____
   (technique of measurement – Sekki disc/glass cylinder)
d) water smell
_____
   (its presence or absence, description and strength)
11. Characteristics of growth on underwater objects (periphyton) _____

(color, shape and size)

12. Water surface pollution _____

(oil film, spots, foam, different floating objects, algae gatherings, etc. and their extend)

13. Fauna of the water body and its vicinity _____

(water invertebrates, insects flying above the water surface, fish, birds, etc.)

14. Main types of human impact _____

_____

_____

(industrial, household and agricultural sources of pollution: presence, intensity and distance from the studied area)

Authors of the description _____

Integrated Study Based on Landscape Profile

This manual continues the study of landscape components along

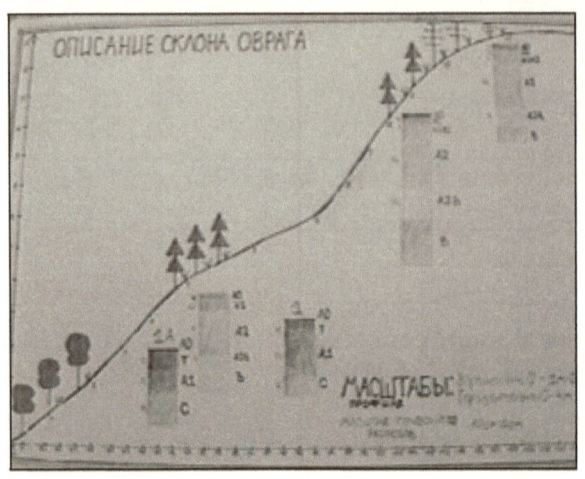

the line of a complex landscape profile on the slope of a river or stream valley. The procedure for a standard geobotanical description of a vegetative cover as well as general rules for a soil profile and its description are provided. The lesson is aimed at describing plant associations and underlying soils and their interrelations.

Introduction

An important aspect of the study of ecological relationships is the determination of spatial distribution of ecosystem components within the borders of the given territory. The main task is to reveal interrelations between vegetation and its environment. Such data can be obtained by two major methods: by making a **geobotanical map** or as a result of laying an integrated **landscape profile**.

The lesson is **aimed** at describing all plant associations and underlying soils along the landscape profile, as well as **revealing their interrelations**.

This study represents a continuation of activities based on laying a profile of a river valley slope with the help of a leveling survey (Lesson 3, fall). Remember that the earlier lesson should result in a relief profile of a small river valley slope at a site of 100-300 meters long with a difference in attitude from 5 up to 15 meters. With the present lesson, students will study all plant associations found along the profile line and soils underlying these phytocenoses at the same site.

Additional objects of study may be some components of the animal world (for example, invertebrates found in ground litter) and microclimate. Larger landscape components (such as geological foundation, surface waters, vertebrates, etc.) cannot serve as objects for study due to the small size of studied area.

The following items **are required** for completion of this educational task: 50 meter measuring tape, previously drawn profile of relief, forms of geobotanical and soil descriptions (one form for each plant association) and spades.

Laying a Landscape Profile

Information on the procedure of laying down a complex landscape profile is given in the manual for lesson 3. Here we will only recall the **main rules** of choosing the site for studies.

1. It is advised to choose a site which is not just a slope of a hill or a mountain, but is a part of the slope of a water-course valley. Ideally, such a site should include such parts of the valley as: a riverbed, a river channel bank, high floodlands, low floodlands (for example, a plant-filled old bed of the river), a slope of terrace over the floodplain, one or several terraces, watershed slope, watershed.

2. It is recommended to choose a site with a slope of about 100-300 meters long with difference in elevation from 5 up to 15 meters. Study of a shorter site or a site with a different slope cannot reveal spatial

differences in landscape components; whereas studies of a longer site will be too laborious.

3. It is recommended to make a profile line, along which the leveling survey is carried out and the soil and geobotanical research will be conducted: a straight line, so that the whole slope is well observed. **A thoroughly planned** profile will allow students to trace plant association distribution depending on environmental conditions, to understand ecological peculiarities of plant associations found along the profile, and to carry out their comparative analysis. If the profile

has not been laid in advance and the area of study is flat, it is recommended to choose a small site (300-500 meters long) with as many different types of vegetation as possible, with different degrees of wetness or different soil types. It is always possible to find such a site in any river valley. In any case it is recommended to pick out a site for studies **within a river valley**.

Geobotanical Research at the Profile

Geobotanical studies at the profile are aimed at identification of plant associations and their boundaries along the profile line; geobotanical description of these plant associations and their mapping on a scheme of the area (in our case, on the profile (section) of relief). Geobotanical study starts with the **determination of the composition of plant associations** found along the profile line. It is the most "intellectual" part of work, requiring from students not only the knowledge of certain plant species, but also familiarization with typical phytocenosis or, conversely, determination of only one plant association (for example, "forest"). Determining plant associations can be quite subjective, especially for beginners. In order to make the performance of this initial stage of work easier for students, a teacher can show and name plant associations along the profile line.

In a typical case, i.e. when studying a profile along the river valley slope from the riverbed up to the watershed slope, researchers can find **5-10 plant associations**. They can include, for example, the following plant associations: flood plain meadow (of 2-3 types depending on moisture conditions and composition of plants), floodplain (valley) forest (more often it is a deciduous forest, for example, alder thicket) or scrubs, a transition zone ("edge": shrubs + deciduous trees + coniferous trees), 2-3 types of forest along the line from the lower terrace up to watershed (for example, fir-wood, which gradually turns into a spruce-and-pine forest, which turns into pine forest).

When determining a phytocenoses at a profile, it is important not to go "too deep" into identification of "microcommunities." Only major, "contrasting" plant associations which are easily recognized visually along the profile line, should be determined.

The next stage of studies consists in **marking plant association boundaries**. The work should begin by familiarizing oneself by passing along the profile line in different directions twice or three times. At the same time, it is necessary to identify and mark boundaries of different phytocenosis at the site and then to determine the length of each plant association along the profile line. In order to make further field work easier and for more clarity, it is recommended to mark boundaries of phytocenosis with pieces of paper, which can be fixed onto tree trunks or pinned to tree branches or on poles driven into the ground. The distances between boundaries of phytocenosis are measured on site with a measuring tape or by counting steps, and are plotted on a prepared map-scheme (in the simplest case – on a line, drawn on the piece of paper) in any chosen scale.

The next stage is to **conduct a geobotanical description** in each of the found plant associations, which are described according to a standard procedure using a table with columns for each feature of vegetation description (see Table 1 Forest Vegetation Description Form in the "Assessment of the vital state of coniferous underbrush" manual, this series, above). Forms are filled in directly on site, where the description is carried out.

Filling in the Report Form

General information on the subsequent description and location of the described plant association including **date, names of authors,** and **number of description** on the profile are written on the form (see Table 1).

Administrative position and location (region, district, position with regard to nearest human settlement) should be described for each plant association. Location of a description point on the profile is also described, for instance, *it is located at the distance of 80 meters up from the water level; it is situated 10 meters eastward from the central line of the profile.*

The position in relief is the determination of location of the given phytocenosis on the profile, for instance, *a floodplain slope, a slope of the first floodplain terrace* and so on.

Environment: all plant associations, surrounding the given phytocenoses, are named here, for example, *a floodplain meadow is situated down the slope, whereas a pine forest is found up the slope.*

The described area (m x m) is the size of the site where description is carried out. Usually sites about 10 x 10 or 20 x 20 are chosen for description of forest associations, whereas sites with size of 1m x1m are selected for description of meadows.

Name of plant association: The name of a plant community is determined by the dominant plant species (or ecological group) from each layer of the phytocoenose. Thus the species within the limits of the same layer are listed **in ascending order** of their total numbers. The full name of forest phytocoenose includes four basic components of vegetative cover: a forest layer, a bush layer, a moss-lichen layer and a herbaceous-shrub layer. They are listed in the exactly same order as described here, for example: *a birch-pine forest with understory consisting of spruce, filbert-mountain ash, and pleutropous bilberry – small-reed.* It is a forest where pine and birch dominate in the arboreal layer (more pines, fewer birches), mountain ash and filbert dominate in the bush layer (mountain ashes prevail), the moss Pleurozium schreberii dominates in the moss layer, small-reed prevails in the herbaceous-shrub layer with less (or as much) of bilberry.

Sometimes, depending on the purpose of the description, a simplified name of the forest can be given, listing main ecological groups of plants that from the phytocoenose, for example: *a birch-pine green moss-mixed herbaceous forest.* In this forest, pine and birch prevail in the arboreal layer; an ecological group of green mosses (various species) dominates in the moss-lichen cover, and cereals and meadow plants of rich soils – in the herbaceous-shrub cover.

The forests with developed moss-lichen cover are usually subdivided into three types corresponding to the prevailing ecological plant groups of this layer: *white mosses* (with a cover of lichens), *long mosses* (with a cover of sphagnum and hair-cap mosses) and *green mosses* (with a cover of Pleurozium and Hylocomium etc.).

Meadow communities are named according to dominant species; the name should also reflect the position of plant associations within relief (upper or lower floodplains, bottomland or dry meadows, etc.). It should be remembered that the name given by a researcher is conventional and thus generic, thus it does not characterize the mentioned plant association in full. The name is given in order to make subsequent analysis easier, so it should not be too long.

The description of arboreal and bush layers

After filling in a cap of the form (general information on the biotope), **the description of arboreal and bush layers** follows. It includes determination of tree crown density, formulas of the forest stand, diameter and height of trunks, height of the attachment of crowns (the lower crown's edge) and age of trees (see Table 1 in the "Assessment of the vital state of coniferous underbrush" manual, above).

For the purposes of this study (a study of layered structure of the forest will be further discussed) the parameters of crown density and formulas of forest stand should be estimated **separately** for each forest canopy: for adult forest stand, young growth and understory.

Crown density

The description of a layer should start with estimation of crown density. Crown density is a portion of the ground surface area covered with crown projections. It can be also characterized as the part of the sky that is closed in by crowns. In other words, the ratio between "open sky" and crowns should be estimated.

Density, abundance and other similar values in geobotany are usually evaluated with the help of one of the three parameters: percentage (from 0 up to 100), in numbers or points (from 1 up to 10) and in fractions of a unit (from 0.1 up to 1), which are, in fact, the same.

Density of crowns is usually expressed in a fraction of a unit - from 0.1 up to 1, i.e. the absence of crowns is taken as zero, and a complete cover of crowns - is 1. Thus, openings among branches are not taken into account; a "crown" is space outlined mentally along

external branches, or the perimeter, of the crown.

It then follows that a dense birch forest in winter, although looking completely "transparent" at an upward glance, in fact, at closer

examination appears to be at maximum density (up to 1). A good method for determining a deciduous forest density when there are no leaves on the trees is to imagine this forest in summer, with all its foliage.

After evaluating species composition and crown density of the arboreal layer, the same parameters should be estimated for **young growth and understory**.

Young growth is comprised of young trees of the main forest-forming species of the given forest up to 1/3 high of the main canopy. Young growth is allocated as a separate canopy of the arboreal layer.

The ***understory*** consists of arboreal and bush species, which can never form a forest stand.

A typical example of young growth in a pine-spruce forest are young spruces, pines, and birches, whereas an understory may be willows, mountain ash, buckthorns, raspberry etc.

It is always a little bit more difficult to determine the density of the

young growth and understory due to their small height and because it is not always possible "to look against the light" from below. In order to learn how to use the procedure of crown density

evaluation for young growth and understory, a reverse psychological trick can be applied: density is estimated as a projection of crowns on the ground; in this case one should imagine, for example, what shadow that the crowns of low trees and bushes will produce and what percentage of the ground would be covered with this shadow.

Crown density should be estimated for each defined layer and canopy of the forest separately: ripe forest stand, young growth and understory.

It is easier to learn how to do this when imagining that there are no other layers or canopies in the forest except the one at hand, and then trying to estimate crown density of this layer alone. Proceed with following layer in the same fashion. It should be noted that in complex multi-layered forests, the total density of crowns of various layers can be more than 1 (due to overlapping crowns in different layers).

Formula of the forest stand

After crown density is estimated, a formula of the forest stand should be composed. It is an estimate of the share (fraction) taken up by a separate species in arboreal and bush layers.

In forest geobotany a share of various trees is determined by a ration of trunks. The share of each species in the formula of the forest is usually expressed in points - from 1 up to 10. The total number of trunks of all plants is considered to be equal to 10, and then the estimations are calculated: what part (share) each separate species takes. If the representation of autonomous (separately growing) plants in the forest is less than 10 % (less than 1 point), they are marked in the formula with a "+" sign, whereas individual plants (1-2 specimens within the studied site) are marked with the sign "ind."

Names of species in the forest formula are reduced to one or two letters, for example: birch - B, oak - O, pine - P, spruce - S, aspen - As, speckled alder – SA, European alder - EA, linden - Ln, larch - La, buckthorn - Bt, raspberry – Rs, filbert – F, etc.

Examples of the formulas of the canopy of an adult forest stand:

1) The formula 6S4B means that the ripe forest stand is formed with 60% spruce and 40 % birch.

2) The formula 10S means that the forest is homogeneous, consisting of only one species - spruce.

3) The formula 10S+B means that in the forest stand there is a minor admixture of birches in the spruce forest.

Taking into account the importance of crown density evaluations as well as the formulas for each of the forest canopies, the records in the description form can, for example, look as follows:

Arboreal and bush layers	Crown density	Formula
Ripe forest stand	0.8	6S 2P 2B
Young growth	0.3	10S
Understory	0.1	5Bt 5 F + Rs

These records mean in the described forest there is a dense canopy of ripe trees. Eighty percent of the space in the top part of the forest is occupied with crowns. Thus, spruce prevails, fewer pines and birches are found, and in equal quantity to each other. There is a rather dense young growth of spruce (an intensive reforestation). Understory is thinned out and consists of buckthorn and filbert at approximately equal shares with separate sprinklings of raspberry. Description of arboreal and bush layers also includes such important information about their structure as the diameter of trunks (D 1.3), height of the forest stand (Hfs), height of crown attachment (Hca) and age of plants.

The **diameter** (D) is measured for several tree trunks that are typical for the given forest at chest height (~1.3m), and then an average diameter is calculated. If necessary it is possible to mark minimum

and maximum values for each canopy as well. The measurements are taken either with a special tree caliper (large sliding calipers), or are based on the length of its circumference. For this purpose, the circumferences of all trees in the site are measured, and then an average value is calculated. The formula $D = c/\pi$, where D is diameter, c is circumference, and π is a constant, equals approximately 3.14 (in field conditions the length of a circle can be

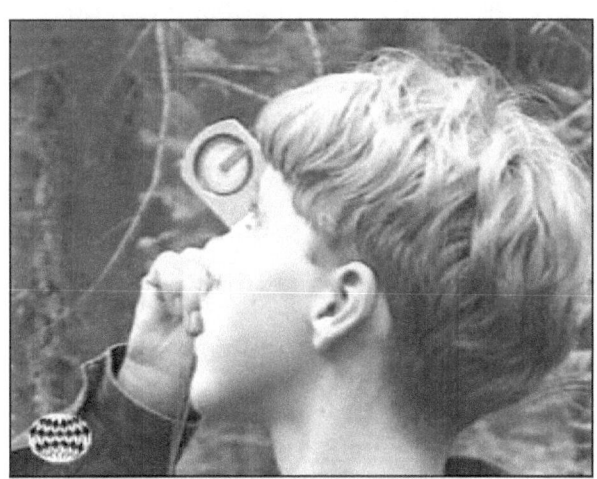

simply divided by three). **Height of the forest stand** (Hfs) is an average height of trees in each separate layer. The measurement of height is carried out according to one of five methods: 1) using an

altimeter – a special optical instrument, which measures a vertical angle at the object with subsequent estimation of the distance from the instrument to the base of the object (in this case, a tree), 2) **by**

eye (which requires experience), 3) using a **tape measure** for measuring of one of the fallen trees of the given canopy, 4) the method of "counting people" and 5) the method of shadow measurement. The first three methods do not require any explanation.

The method of **"counting people"** consists of the following. Measurements are carried out by two people together: one person stands close to the tree, and another, who is skilled at visual estimation, walks from the tree until he or she can cover the whole tree from the base up to the top in view. Then he or she tries to count how many people of the given height can be "stacked" along the trunk.

It is recommended each time to postpone distance, which is twice longer, than the previous one, i.e. while "moving" mentally along the tree trunk upwards, first, height of two "person" should be postponed, then height of two should be added, then – of four, then of eight etc. (i.e. according to the scheme: 1 - 2 - 4 - 8 -16). From the point of view of human visual estimation it is easier and more accurate. Knowing this "person's" height it is possible to calculate the height of the tree. The fifth method is the most precise of all indirect methods, and is used in sunny weather. The **shadow** from a standing person whose height is known is measured precisely. The shadow from the studied tree is then also measured. In the dense forest, when the shadow of the given tree and, especially, its top is difficult to find, the following method can be employed. A person should walk from the tree so that his or her sight (head), the top of the tree, and the sun lie along one line. Then the person should find a shadow of his or her own head on the ground – which will also be a shadow of the top of the tree. Then only the distance between this point and the base of the tree should be measured. The height of the tree is estimated according to a proportion: length of the shadow of the person / its height - length of shadow of the tree / its height.

For the educational purposes it is possible to use a combination of several methods – so that measurements received by direct methods will verify results of measurements by indirect methods.

The **height of crown attachment** (H ca) is an average height at which lower living branches of trees are found (it is not recorded for the young growth or the understory).

Age of trees can be more reliably estimated according to the annual rings of cut trees, which can be found practically in any forest (of course it is not necessary to cut down trees for this purpose). Some recently cut trees or their stumps can be used for these purposes. If there no "fresh" stumps in the forest it is necessary to make a complete cut cross-section or to cut down a trunk of the fallen tree with an axe - at least up to its core. The cut cross-section should be taken out of the tree as close to its butt end as possible.

The age of the young growth is also defined according to annual rings using the example of one sawed or cut down plant (it is better to do this outside the studied site).

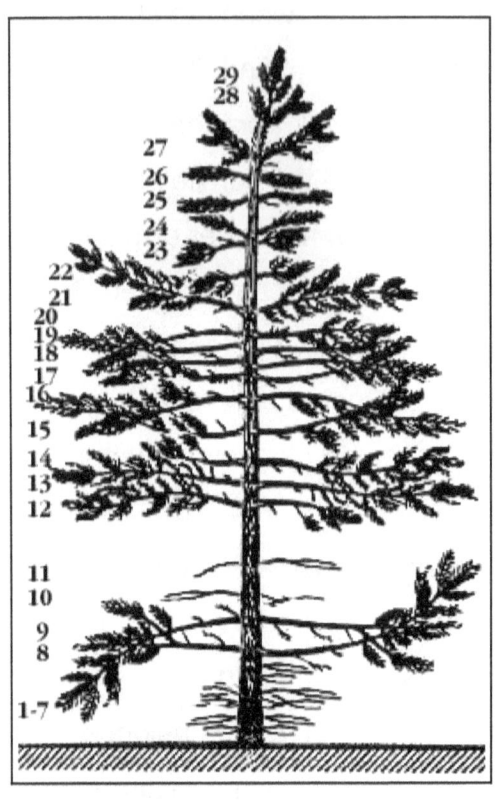

The age of young and middle-aged trees, spruces and pines in particular, can be defined according to whorls, which are several branches at the same level around the trunk along all its length. These plants still have dead (in the bottom part of a crown) or living (in the top part) branches, which grow by bundles – whorls -- until they are 30-40 years old (and sometimes even longer. The quantity of such whorls - from the base of the trunk up to its top, precisely corresponds to age of the tree, because during one vegetative season a tree grows one internode (one whorl). The number of years arrived at by counting whorls should be increased by at least three years, taking into account the rooting period and the start of growth.

The description of herbaceous-shrub and moss-lichen layers

When the description of the arboreal-bush layer is finished (after filling in the table), then one can start with the description of the herbaceous-shrub and moss-lichen layers.

The form of the description of this section of vegetation provides for the presence of different forms of microrelief – hillock (i.e. a small rise), and the space between (i.e., a depression), which usually differ among themselves according to their species structure and plant distribution (Table 1 see in the "Assessment of the vital state of coniferous underbrush" manual, above). If such forms of a microrelief within a described site are not present (the ground surface is flat, plain), then the description of herbaceous-shrub and moss-lichen layers can be written in one column, and subtitles of the microrelief can be simply deleted.

In fact, **the description of herbaceous-shrub layer** includes compiling a list of plant species at the given site with rough estimations of their abundance. The determination of plants should be carried out using all available means: with the help of scientific guides, atlases, guiding tables, more experienced students, a teacher. There is no sense in determining absolutely every herbaceous plant species within the site, because it is too complicated and is not necessary for the purposes of this lesson. It is important to define only ten main or dominant herbaceous plant species on the site as well as their relative numbers.

A simple indicator of an herbaceous plant's abundance is the value of *projective cover*. Projective (or descriptive) cover for herbaceous plants is actually the same parameter as crown density for arboreal and bush layers.

Projective cover is expressed as a percentage and is determined separately for each species. Thus the sum of the projective cover values of all the species can be more than 100 (in rare cases) if the leaves of plants are overlapping (form several "canopies"). If an area of the soil's surface is left uncovered by plants, then the total projective cover can be less than 100%. The accuracy of projective cover estimations should be not less than 5 %.

When filling in the column "herbaceous-shrub layer" of the form, the names of plants have to be written in one column, or in several, if the list of plant species is quite long. It is recommended to first list shrubs (bilberry, red whortleberry, etc), and then herbaceous plants in decreasing order according to number (projective cover). Rare plants, with projective cover less than 5 % are consolidated with brackets and their total projective cover is given within them. Individual plants, as in the case of an arboreal-bush layer, are marked with the sign, "ind."

Then the **moss-lichen layer** is described in the same way as the herbaceous-shrub layer, listing the names of found mosses and lichens (if they are found on the ground and their determination is possible) as well as recording the projective cover for each species. Determination of mosses and lichens as well as herbaceous plants is rather difficult so it is enough to determine the first three to five of the most numerous species.

Unknown plant species found during the description are sampled for the herbarium and taken home for further study. Thus they should be given a certain number (index) in the description form, which is replaced with a species name after determination. The procedure of geobotanical description and filling in the form is completed for each of the plant associations found along the profile line – from its bottom up to the top.

When the task is performed by a group of 10-15 students, it should be divided into teams of 2-3 people. It is recommended that you select **contrasting** plant associations, for instance, one meadow community and one forested site, to be described by the same team of students.

Soil studies at the profile

The next stage of field studies is to **lay soil sections in each plant association** found along the profile line.

The procedure of preparing a description of a soil profile (section) is given in full detail in the "Simple Procedure of Soil Description" manual, Part 1: Geography and Landscape Sciences), so we will only mention **the main rules**, which are standard for any soil study.

1. A soil section profile is laid on the most typical site of the studied phytocenosis. It **cannot be laid** down close to roads, ditches, on paths or trampled sites, or at elements of microrelief that are not typical for the area (in a depression, on hummocks, etc.).

2. Ideally a soil pit is dug down the whole depth of the soil, i.e. down to the depth of the underlying bedding mother rock (from 20 up to 120 cm depending on the soil type). The front side of the soil section should face the sun, and it must not be stepped on or damaged; excavated soil is put to the side of the pit. It is necessary to create stairs into the pit. The front side is scraped immediately before the description.

3. Description of soil horizons is carried out according to a standard procedure by filling in the form of soil descriptions, including: 1) general information on the soil section (profile), 2) soil structure (presence of genetic horizons), 3) thickness of soils and same soil horizons, 4) morphological features of horizons: coloring, moisture, mechanical composition, structure, presence of new formations and inclusions.

For each soil section along the profile line, it is recommended to determine functional soil zones (accumulative, eluvial, illuvial and mother rock), to determine their position with regard to each other and their thickness, as well as to name the soil, if possible.

Presentation of results

The studied complex landscape profile should be drawn up visually as a report on the conducted research.

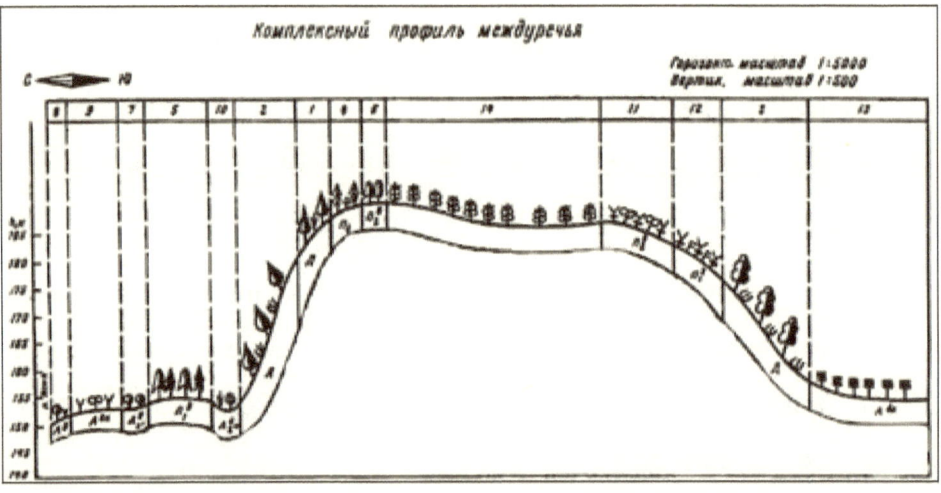

The line of relief plotted on scale or any other square paper forms the basis of the profile. **The scale** of the drawing depends on the form of presentation – if an oral report is prepared, then the drawing can be made on a large sheet of paper or poster. If an article is prepared, then the drawing can be made on normal letter size paper. In any case, the following information should be present on the drawing: **a name** (for example, "Complex landscape profile of the river... valley slope"), **signed axes**, where length and height of the profile are marked, the relief line, vegetation, soils, as well as conventions (map symbols).

Vegetation is drawn on the profile line with symbols representing different wood species, shrubs, and grass vegetation without any

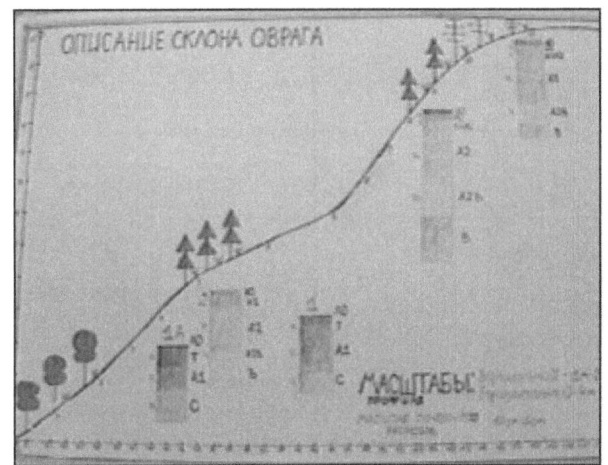

regard to the chosen scale. These symbols can be conventional, but it is advised to make them "look like" natural studied phytocenoses. Thus, the composition of wood species in the forest and the forest

layer structure can be shown by schematic images of trees of different forms. Shrub and grass layers can also be shown as described above. **Boundaries** of phytocenoses along the profile line should be marked with colorful vertical lines drawn in the corresponding places. Names of plant associations can also be written on the drawing.

Soil sections are drawn below the relief line in the form of a colorful scheme – columns within the corresponding plant associations. The following characteristics of soils and soil horizons should be marked on the soil section with regard to the chosen scale (which should be written on the drawing as well): thickness of soil and separate soil horizons, their natural color, structure and inclusions (with the help of conventions), if possible. The **legend with conventions** should be on the drawing separately for vegetation cover and soils.

The following information should be **attached** to the report as an appendix: tables of leveling survey, forms of vegetation description and forms of soil descriptions.

When finalizing a presentation of the final report, it is necessary to make provisions for the possibility of adding results of a **snow-measuring survey** to the studied landscape components. The snow-measuring survey is to be conducted at the end of winter (Winter Lesson №10), and if possible, at the same complex profile.

While interpreting results of the research, it is recommended to answer the following questions:

1) Is there a spatial connection among the three studied components of landscape: relief, soils and vegetation?

2) What plant associations correspond with what relief forms and why?

3) What factors of physical environment influenced the formation of certain plant associations at certain relief sites?

4) Are there any differences in the soils that underlie different plant associations?

5) What factor has the most influence on soil: vegetation or location in relief? What can this influence consist of?

6) What influence does modern relief have on the interrelation between soils and vegetation?

Study of growth dynamics of trees based on annual rings

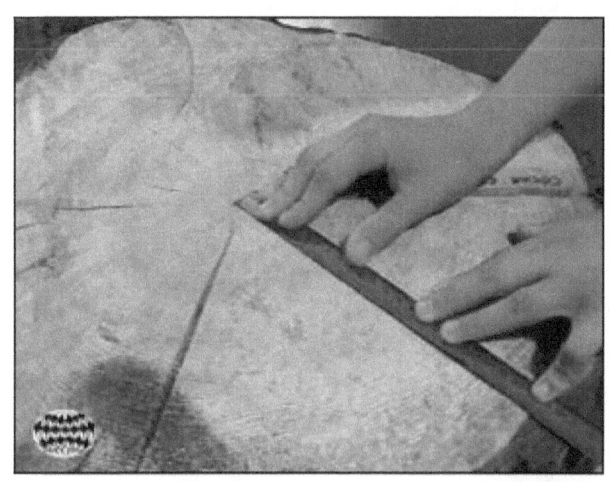

This manual describes an elementary procedure for the preparation of a tree trunk cross-section and subsequent counting of annual growth rings. The information will then be used to sketch a graph of the dynamics of tree growth in years and further analyze the tree growth in connection with changes in environmental factors.

Introduction

The given educational task is devoted to the determination of the **age** of a tree based on the number of annual rings and the study of peculiarities of tree **growth** during various years based on the width of these rings.

The annual rings seen in a cross-section cutting of a tree trunk of a tree grown in a moderate climatic zone occur as a result of the varying rates of growth during a vegetative season. The cells that are formed in the spring and summer have a lighter tone. The cells that are formed at the end of the vegetative season form wood that is composed of smaller cells whose cell walls are thicker than the ones formed in the spring and summer. The color of these smaller cells is darker than those formed in the beginning of the summer. Thus, an annual ring has light and dark components, and as a result, we can see borders of annual rings on the cut cross-section of the tree.

This occurs only in those zones of the earth where there is a

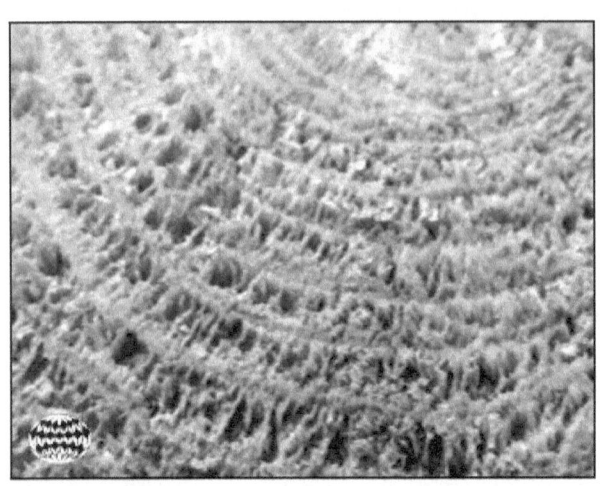

noticeable change of seasons. In regions without a change of seasons, for example on the equator, annual rings are also formed, but they are practically invisible - wood has a even coloring.

By looking at the number of annual rings on the cut cross-section of a trunk it is possible quite **accurately to define the age of the tree**. The width of one annual ring, i.e. annual growth ring, varies from one year to the next. The width depends on tree state (condition) during the given vegetative season, which, in turn, depends on annual climatic peculiarities (features), health of the tree and many other factors.

The average width of rings of different tree species varies and is basically associated with the species, the place it is grown, and individual features of the given tree. However, there are some typical features for all tree species or for the majority of trees of the same species. For example, rings are wider on the well-illuminated side of the tree than on the side of the tree that is in the shadow, therefore the stumps that remain from lone standing (grown) trees are used for the determination of Cardinal points (where North and the South are).

Within a species beginning to grow under the forest canopy (in our woods they are, for example, spruce and oak) it is possible to observe objective changes of ring width during the first years of the trees life: while the young tree is growing in the shade, rings are narrow, when it begins getting more sunlight, rings become wider.

Natural loss (fallout) and the cutting of neighboring trees during forest management (maintenance) can also influence the width of the rings. When a "window" (opening) is formed near a growing tree, it begins to actively grow, growing both in height and in thickness.

The goal of this lesson is the preparation of a tree cross-section and the construction of a graph of its growth. Interesting results can be determined if students prepare cross-sections of several trees, growing in one general area but under different environmental conditions (for example different habitats) and then comparing the graphs.

We recommend doing this laboratory-based activity as part of winter ecological field studies during cold weather, while other outdoor field studies may be difficult and uncomfortable to execute.

For this lesson it is **necessary to have**: two-handled saw and metal ruler. For preparation of a permanent cross-section for multiple years, it is necessary to have a metal strip and nails.

Preparation of the cut cross-section

The study of growth dynamics of a tree can be conducted at any time of year and without dependence on weather conditions. For conducting the study, a saw, a ruler, a pen, in some cases a magnifying glass and some painting substance such as bluing or potassium permanganate are required (if no prepared cross-sections are available).

A wind-fallen tree or a standing dead tree should be found in the forest, whose year of dying off can be defined. If the needles still remain on a tree, then the current year can be taken for the year that it died. If there are no needles, but the smallest branches are present – the tree died the previous year. If there are no needles and there are no small-sized branches present, but the cortex is well preserved – it died about two years ago. It is advisable not to use older trees as it is impossible to define the year of the tree's death precisely, so all further attempts will produce no results due to absence of the "starting point" of chronological scale.

An ideal tree for the preparation of cut cross-section is **a recently wind fallen tree**. Its death is not caused by natural reasons, i.e. by illnesses or pests, but by the influence of external forces. The analysis of such trees gives good results on the growth dynamics of the tree during its last years.

Cut the cross-section of the trunk as close to the base of the tree as possible. This is recommended for a more accurate estimation of the year of tree's birth. In any case, at the subsequent estimation of the year of tree's birth, a number of years are added to the calculated age of a trunk at the level of cutting. This number corresponds to the years that the tree has grown up to the height of cutting. Half a meter distance corresponds approximately to 5-7 years, about one meter, 10-12. The cut cross-section is made with a hand-held or power saw.

In order to cut an educational cross-section (i.e. not to be used only once but to be stored and subsequently examined many times by students), a disk is prepared, i.e. the trunk is twice sawed. Thus two cross-sections are made at the distance approximately about 10-15 cm from each other. A thinner disk may break apart, whereas a thick one will be too heavy. The disk is stripped in the lab. If the rings are not easily visible, it is possible to **dye** the cross-section, for example, with potassium permanganate, which will make the rings more distinct. For long-term storage and multiple use it is better to **frame** the edge of the disk along the circle with metal tape – it will protect the disk from breaking up when it dries and cracks (which is inevitable).

Estimations of ring width

The most responsible procedure for estimating the width of annual rings is done in the following order:

First, **a line** is sketched with a thin pencil; all measurements will be conducted along this line. The line should pass precisely from the cross-section center to its external edge (along the radius). A sector of the trunk with the least number of anomalies - cracks, not concentric compressions, remains of knags, old leaked wounds etc. should be chosen for measurements. The line of measurements should cross the most "average" sector of wood.

Then a ruler with very distinct millimeter divisions (a metal ruler, for example) is applied to the external edge of last (outside) ring. The zero of the ruler should coincide with the external edge of the last ring. In order to make sure that the ruler cannot be shifted inadvertently while measuring, it is better to press it down with something heavy or to attach it to the wood with pins in several places.

A working table is then prepared for data recording of measurements (example):

Year	Mark	Growth	Year	Mark	Growth	Year	Mark	Growth
1999	0	2.2	1934	154.5	2.7	1878	344.5	5.2
1998	2.2	1.9	1933	157.2	1.8	1877	349.5	5.0
1997	4.1	...	1932	159.0	...	1876	354.5	...
...	...	...	...	...	...	...	...	...

The number of cells in the table should approximately correspond to age of the tree.

In the column "year," all years are recorded in advance, starting with the year of the tree's death, into the course of time, **back to the year**

of the tree's birth, which is precisely defined when the measurements are over.

In the column "mark," readings of the position (location) of the next annual ring's **border** on the ruler are recorded in millimeters. Such measurements are taken consequently from the latest (external) to the very first annual ring (in the center of the cross-section), gradually moving from the cross-section edge to the center and recording each of the following measurements in the appropriate space of the table. Having added 5-10 years (depending on the height the cut was made) to the first year (in the center), the year of the tree's birth can be determined. This value has no principal importance; it can be approximate because the years are counted in the reverse order, starting with the last annual ring, which corresponds to a known year.

After completing the measurement of the rings' position along the radius (after filling in the "mark" column), **calculations of annual growth** should be recorded. They are calculated with the help of a calculator, subtracting the value of each older ring from the value of younger ring's disposition. For example, for the case indicated in the table – the value of 1998 (2.2) is deducted from the value of 1997 (4.1) and we determine annual growth in 1998 as equal to 1.9 mm.

Such estimations are calculated for all years of the given cross-section. The data forms the **basis** for making a graph of the growth dynamics of the tree in years.

Graph plotting

On the basis of the data on annual growth the graph of dynamics of the tree's growth is plotted according to years. It is more convenient

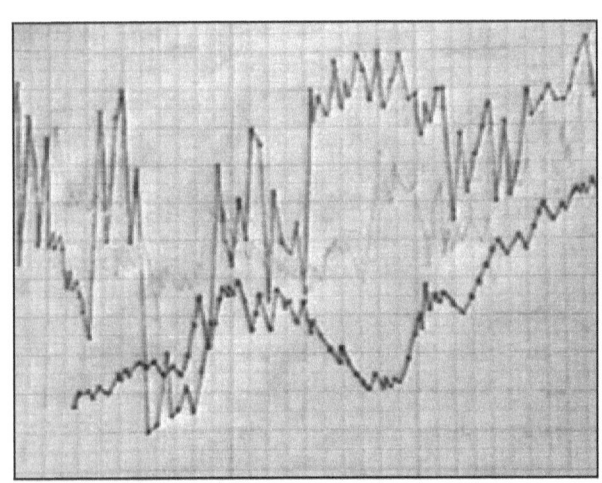

to plot it on squared or, even better - on scale paper.

Years - from left to right starting with the year of the tree's birth to the last year of its life are laid off as abscissas. The years should not be plotted all in a row, because the marks would be too crowded together. Rather, place their marks on an axis of abscissas, one for each 10 years.

Absolute values of annual growth are laid off on the axis of ordinates in millimeters.

The scale should be chosen so that the axis of abscissas is approximately twice longer than the axis of ordinates. Then the graph will be visual and easily readable.

Data interpretation

Analyzing the graph, students should try to define years with **minimum** and **maximum** annual growth, and if possible, to discern **long periods** of the slowed and accelerated growth of the tree. It is necessary to try to connect these declines and rises with some external factors. It is ideal, if possible, to get data on climatic conditions of those years of abnormal growth. Such data is available at meteorological stations. Using their data it is possible to reveal those climatic factors, and to determine which factors had the greatest influence on the growth of trees.

When analyzing the graph, special attention should be paid to the **first and last years** in the life of the tree, as some interesting details can come to light whether the tree lived the first years in favorable conditions or how long it was sick before dying off (if the death was from natural causes). It is also interesting to analyze the influence of sanitary cutting and forest management cuttings on the growth of the tree in the past. For this purpose, it is possible to use the data on cuttings conducted by the local forestry station.

The work will be more interesting if several cut cross-sections are made of trees of different species or grown in different places (if sufficient "labor resources" are available). For example, it is possible to compare the dynamics of tree growth of a spruce and a pine, or pines from a pine forest, a mixed forest, and a swamp. In such cases common (general) regularities of a tree's growth in the region (we shall say which are determined by the climate) and individual features of trees (for example, vital status) are revealed.

The final report on the conducted study should consist of the graph of dynamics of tree growth (or several trees on the same graph) with revealed **periods** of slowed and accelerated growth. The **photos** and **explanatory description** (note) in the text should be attached to the graph, explaining observed facts and regularities (if found).